Lecture Notes in Mathematics

Edited by A. Dold and B. Eckmann

416

Michael Taylor

Pseudo Differential Operators

Springer-Verlag
Berlin · Heidelberg · New York 1974

Dr. Michael E. Taylor
University of Michigan
Ann Arbor, MI 48104/USA

Library of Congress Cataloging in Publication Data

Taylor, Michael Eugene, 1946-
Pseudo differential operators.

(Lecture notes in mathematics ; 416)
Bibliography: p.
Includes index.
1. Differential equations, Partial. 2. Pseudo-differential operators. I. Title. II. Series: Lecture notes in mathematics (Berlin) ; 416.
QA3.L28 no. 416 [QA374] 510'.8s [515'.724] 74-23846

AMS Subject Classifications (1970): 35-02, 35S05

ISBN 3-540-06961-5 Springer-Verlag Berlin · Heidelberg · New York
ISBN 0-387-06961-5 Springer-Verlag New York · Heidelberg · Berlin

Offsetdruck: Julius Beltz, Hemsbach/Bergstr.

TABLE OF CONTENTS

LECTURES ON PSEUDO DIFFERENTIAL OPERATORS

INTRODUCTION: These notes are based on the lectures I gave in partial differential equations at the University of Michigan during the winter semester of 1972, with some extensions. References to further work have been added at the end of Chapters III, IV, and V, and a few exercises have been thrown in, in addition to those thrown out in class.

The students to whom these lectures were addressed were assumed to have knowledge of elementary functional analysis, the Fourier transform, distribution theory, and Sobolev spaces, and such tools are used without comment. We refer the reader especially to Yosida [85] for the background material. The last section of Chapter 1 also relies on the basic results of C* algebra theory, and the reader who doesn't like functional analysis might have to skip this section on first reading. Beyond that, we have tried to make these notes self-contained.

This is not to say that these notes constitute a self contained introduction to the subject of partial differential equations, and the beginning student would have to see the material in several of the books we have mentioned in the references, especially, [13], [25], [31], and [1], [52], [54], [55], [59], in order to get a good idea of what the subject is about. What we do here is develop one tool, the calculus of pseudo differential operators, and apply it to several of the main problems of partial differential equations.

We begin in Chapter I by describing the earliest sort of singular integrals on the circle investigated by Poincaré, Hilbert, and others, and an application to the oblique derivative problem on the disc. In the second Chapter we introduce the modern calculus of pseudo differential operators, developed by Kohn and Nirenberg, Lax, Hörmander, Kumano-Go, and others, and in Chapter III we apply these results to obtain interior regularity results for elliptic and hypoelliptic operators. The next two chapters are devoted to the main topics of classical PDE, the initial value problem for hyperbolic and parabolic equations, and boundary value problems for elliptic equations. We give a unified treatment of these topics, and Gårding's inequality plays a crucial role here in passing from formal properties of symbols to the energy inequalities and other _a priori_ inequalities needed for various results on existence and regularity.

In Chapter VI we cover some recent work of Hörmander on wave front sets and the propagation of singularities of solutions to partial differential equations. Applications are given to local existence of solutions to PDE's and to an exponential decay result. The proof of the main result on propagation of singularities requires the sharp Gårding inequality which, following Kumano-Go, we prove in the last chapter of these notes.

One important topic we have not included is Uniqueness in the Cauchy problem. We recommend that the reader consult [91].

It is a pleasure to thank Eric Bedford, whose classroom notes greatly aided the preparation of these notes, and Professor Jeff Rauch for some interesting conversations, especially relating to hyperbolic equations.

CHAPTER I. SINGULAR INTEGRAL OPERATORS ON THE CIRCLE

The basic singular integral operator with which we will be concerned here can be described as follows.

If $u \in L^2(S^1)$, write $u(\phi) = \sum_{n=-\infty}^{\infty} a_n e^{in\phi}$, and define

$$Pu = \sum_{n=0}^{\infty} a_n e^{in\phi}$$

In order to interpret the operator P, which is clearly a continuous orthogonal projection on $L^2(S^1)$, as a singular integral operator, consider the Cauchy integral

$$Tu(z) = \frac{1}{2\pi i} \int_{S^1} \frac{u(\zeta)}{\zeta - z} d\zeta \qquad (|z| < 1)$$

$$= \frac{1}{2\pi} \int_{-\pi}^{\pi} \frac{u(\phi)}{e^{i\phi} - z} e^{i\phi} d\phi$$

We can rewrite this as

$$Tu(re^{i\theta}) = \frac{1}{2\pi} \int_{-\pi}^{\pi} \frac{u(\phi)}{1 - re^{i(\theta-\phi)}} d\phi$$

$$= \frac{1}{2\pi} \int_{-\pi}^{\pi} \frac{u(\phi) - u(\theta)}{1 - re^{i(\theta-\phi)}} d\phi + u(\theta)$$

If $u \in C^1(S^1)$, we can pass to the limit as $r \to 1$ and obtain

$$\lim T u(re^{i\theta}) = \frac{1}{2\pi} \int_{-\pi}^{\pi} \frac{u(\phi) - u(\theta)}{1 - e^{i(\theta-\phi)}} d\phi + u(\theta)$$

$$= \lim_{\varepsilon\to 0} \frac{1}{2\pi} \int_{S^1\setminus I_\varepsilon(\theta)} \frac{u(\phi) - u(\theta)}{1-e^{i(\theta-\phi)}} d\phi + u(\theta)$$

$$= \lim_{\varepsilon\to 0} \frac{1}{2\pi i} \int_{S^1\setminus I_\varepsilon(\theta)} \frac{u(\zeta) - u(\theta)}{\zeta-e^{\theta}} d\zeta + u(\theta)$$

$$= \frac{1}{2\pi i} \text{PV} \int_{S^1} \frac{u(\zeta)}{\zeta-e^{i\theta}} d\zeta + \frac{1}{2} u(\theta)$$

where $I_\varepsilon(\theta) = (\theta - \varepsilon, \theta + \varepsilon)$.
Since it is easy to verify that $\lim_{r\uparrow 1} Tu(re^{i\theta}) = Pu$ for $u \in C^1(S^1)$,*

it follows that $Pu = \frac{1}{2} Hu + \frac{1}{2} u$ where

$$Hu(e^{i\theta}) = \frac{1}{\pi i} \text{PV} \int_{S^1} \frac{u(\zeta)}{\zeta-e^{i\theta}} d\zeta$$

The singular integral operator H is called the Hilbert transform. The formula we have just derived shows that H extends to a continuous linear operator on $L^2(S^1)$.

Exercise 1. Find the Fourier series representation of H. Prove that H is a unitary operator on $L^2(S^1)$ and $H^2 = I$.

* (using the residue theorem, the reader should check that, if $u_k = e^{ik\theta}$, then $\lim Tu_k(re^{i\theta}) = u_k$ if $k \geq 0$, 0 if $k < 0$.)

§1. The algebra of singular integral operators.

Definition: The algebra $\mathfrak{A}$ of singular integral operators on S^1 is the norm closed algebra of operators on $L^2(S^1)$ generated by:

(1) P

(2) multiplication by $a \in C(S^1)$

(3) $\mathcal{C}$, the set of compact operators

Actually (3) is redundant, but we shall not prove this fact, nor make use of it.

Theorem 1: If $A, B \in \mathfrak{A}$, then $[A,B] = AB - BA \in \mathcal{C}$.

Proof: It suffices to show that $aP - PA \in \mathcal{C}$ if $a \in C(S^1)$. Suppose that $a = e^{im\phi}$, $f = \sum_{n=-\infty}^{\infty} a_n e^{in\phi}$.

Then $aPf = e^{im\phi} \sum_{n=0}^{\infty} a_n e^{in\phi} = \sum_{n=m}^{\infty} a_{n-m} e^{in\phi}$, and

$$Paf = P\left(\sum_{n=-\infty}^{\infty} a_{n-m} e^{in\phi}\right) = \sum_{n=0}^{\infty} a_{n-m} e^{in\phi} .$$

The $[a,P]f = \sum_{n=0}^{m-1} a_{n-m} e^{in\phi}$. Hence $[a,P]$ in an operator with finite dimensional range, and therefore is compact. Since trigonometric polynomials are dense in $C(S^1)$, the result holds for all $a \in C(S^1)$, because $\mathcal{C}$ is norm closed.

This theorem, which says that $\mathfrak{A}$ is commutative, modulo $\mathcal{C}$ is important in that it enables us to give a nice conditon that an operator in $\mathfrak{A}$ be Fredholm. For the moment, consider an operator $T \in \mathfrak{A}$ of the form

$$T = aP + b(1-P) + K, \quad K \in \mathcal{C}$$

In section 3 we shall show that every $T \in \mathfrak{A}$ is of this form, but we won't need this, since all singular integral operators one encounters are automatically constructed in this form. For such a T, we tentatively define the symbol σ_T of T, as a function on $S^1 \times \mathbb{Z}_2$ by

$$\sigma_T(\phi,1) = a(\phi)$$

$$\sigma_T(\phi,-1) = b(\phi)$$

We show that σ_T is indeed well defined.

Lemma 1: If $aP + b(1-P) \in \mathcal{C}$, then $a \equiv b \equiv 0$.

Proof: $(aP + b(1-P))P = aP \in \mathcal{C}$, since $P^2 = P$.
Then $|a|^2 P \in \mathcal{C}$.
If $U_\phi f(\theta) = f(\phi-\theta)$, the map $\phi \to U_\phi |a|^2 PU_{-\phi}$ is continuous in the uniform operator topology, since

$$U_\phi |a|^2 PU_{-\phi}f(\theta) = |a(\theta-\phi)|^2 Pf(\theta) \text{ . Thus}$$

$$\frac{1}{2\pi} \int_0^{2\pi} U_\phi|a|^2 PU_{-\phi}\, d\phi = ||a||_2^2 P \in \mathcal{C}, \text{ which forces } a \equiv 0 .$$

Similary we obtain $b \equiv 0$.

Theorem 2: If $T = aP + b(1-P) + K_1$, and $W = \alpha P + \beta(1-P) + K_2$, then $\sigma_T \sigma_W = \sigma_{TW}$.

Proof: This is immediate from the computation

$$TW = (aP + b(1-P) + K_1)\ (\alpha P + \beta(1-P) + K_2)$$
$$= a\alpha P^2 + b\beta(1-P)^2 + K_3$$
$$= a\alpha P + b\beta(1-P) + K_3\ .$$

Recall that a linear operator $T \in \mathcal{L}(L^2)$ is called Fredholm if

(1) $R(T)$ is closed

(2) $\dim \ker T < \infty$

(3) $\dim \operatorname{coker} T < \infty$.

The reader should also recall the following important result from the Riesz theory of compact operators (see [64], Chap. VII)

Proposition: $T \in \mathcal{L}(L^2)$ is Fredholm if and only if there exists $U \in \mathcal{L}(L^2)$, called a Fredholm inverse of T, such that $TU = I + K_1$, and $UT = I + K_2$, where K_1, K_2 are compact.

The following Fredholm property of singular integral operators is now immediate.

Theorem 3: Let $T = aP + b(1-P) + K$. Then T is Fredholm if σ_T is nowhere vanishing.

Proof. Let $U = \frac{1}{a} P + \frac{1}{b}(1-P)$. Then $\sigma_{TU} = \sigma_{UT} \equiv 1$, so U is a Fredholm inverse of T.

In section 3 we shall show that this conditon on σ_T is also necessary for T to be Fredholm on $L^2(S')$.

The problem of how to define the symbol of a singular integral operator on a multidimensional space took quite some time in being solved. Mikhlin defined a symbol in 1936. This symbol was elucidated by Calderon and Zygmund in their important works in the early 1950's. It was Lax who suggested a Fourier series representation to treat multidimensional singular integrals, and the Fourier integral representation used by Kohn and Nirenberg is the one we shall use in the next chapter.

Exercise: Let $Tf(x) = \frac{1}{\pi i} \, PV \int_{S^1} \frac{a(x,y)}{y-x} f(y)\, dy$, where $a \in C^\infty(S^1 \times S^1)$. Show that

$$Tf = bHf + Kf$$

where $b(x) = a(x,x)$ and $K: H^s \rightarrow H^s$ is compact, for all s.

§2. The oblique derivative problem.

Here we discuss one application of the algebra of singular integrals developed above. For further applications we refer the reader to Mikhlin [58] and Muskhelishvili [60] The problem we consider is the oblique derivative problem for functions harmonic on the disc: given $g \in C(S^1)$, find u harmonic on $B = \{z \in \mathbb{C}: |z| < 1\}$ such that

$$\beta u = a \frac{\partial}{\partial r} u + b \frac{\partial}{\partial \phi} u + c u \Big|_{S^1} = g . \tag{1}$$

The way we handle this is as follows. First, the Dirichlet problem

$$\Delta u = 0 \text{ in } B$$
$$u \Big|_{S^1} = f$$

can be solved explicitly by $u = PIf$, the Poisson integral of f. To derive this Poisson integral representation, write

$$f = \sum_{n=-\infty}^{\infty} a_n e^{in\phi} . \quad \text{Then}$$

$$PIf(re^{i\phi}) = \sum_{n-\infty}^{\infty} a_n r^{|n|} e^{in\phi}$$

$$= \frac{1}{2\pi} \int_{-\pi}^{\pi} \frac{1 - r^2}{1+r^2 - 2r \cos(\phi-\theta)} f(\theta)\, d\theta$$

The reader can verify as an exercise that PIf does the trick.

Exercise 2. Prove that if $s \geq -\frac{1}{2}$ and $f \in H^s(S)$, then $PIf \in H^{s+\frac{1}{2}}(B)$. (On first reading, don't take this exercise too seriously.)

In view of the fact that restriction to S' maps $H^{\tau}(B)$ onto $H^{\tau-\frac{1}{2}}(S^1)$ for $\tau > \frac{1}{2}$, we have the following commutative diagram with $s > 0$.

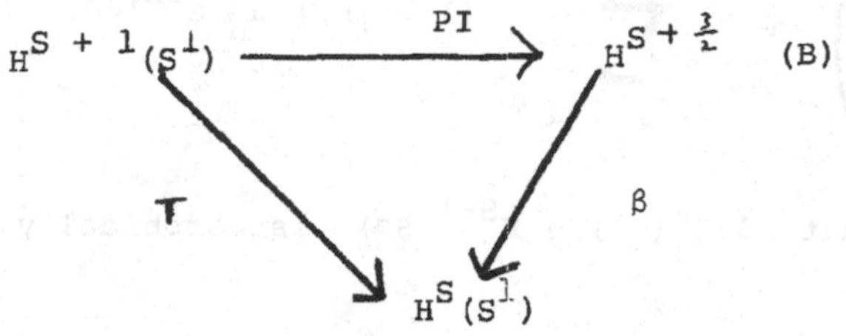

where we define $T = \beta \circ PI$. Hence we can solve (1) by setting $u = PI\,h$ if $Th = g$. Hence solving (1) is equivalent to inverting T .

More generally, we are interested in when problem (1) is Fredholm, in the sense that it can be solved provided g satisfies a certain finite number of linear conditions, and the set of u satisfying $\Delta u = 0$, $\beta u = 0$ should be finite dimensional.

This is equivalent to asking when is T Fredholm, which is right up our alley, since we will now write T as a pseudo differential operator. We have

$$u = PIf = \sum_{n-\infty}^{\infty} a_n \, r^{|n|} e^{in\phi}$$

$$\frac{\partial}{\partial\phi} u \Big|_{S^1} = \frac{\partial}{\partial\phi} f = \sum_{n-\infty}^{\infty} in \, a_n \, e^{in\phi}$$

and $$\frac{\partial}{\partial r} u \Big|_{S'} = \sum_{n=-\infty}^{\infty} |n| \, a_n e^{in\phi}$$

Now let us define an operator Λ by

$$\Lambda\left(\sum_{n=-\infty}^{\infty} a_n e^{in\phi}\right) = \sum_{n=-\infty}^{\infty} (1+|n|) \, a_n e^{in\phi} .$$

Exercise 3. Prove that $\Lambda : H^S(S^1) \to H^{S-1}(S^1)$ isomorphically, for each real S .

It follows that $T = a \frac{\partial}{\partial r} + b \frac{\partial}{\partial \phi} + C = a(\Lambda - 1) + b \frac{\partial}{\partial \phi} + C$

$$= (a + b \frac{\partial}{\partial \phi} \Lambda^{-1} + (c-a)\Lambda^{-1})\Lambda .$$

Since Λ is an isomorphism of $H^{S+1}(S^1)$ onto $H^S(S^1)$ it follows that $T : H^{S+1}(S^1) \to H^S(S^1)$ is Fredholm if and only if $S = (a + b \frac{\partial}{\partial \phi} \Lambda^{-1} + (c-a)\Lambda^{-1}) : H^S \to H^S$ is Fredholm.

Exercise 4. $\Lambda^{-1} : H^S \to H^S$ is compact, for each real S.

Now we compute that $\frac{\partial}{\partial \phi} \Lambda^{-1} (\sum a_n e^{in\phi})$

$$= i \sum \frac{n}{1+|n|} a_n e^{in\phi} = i(2P-1)(\sum a_n e^{in\phi}) - \sum \frac{a_n}{1+|n|} e^{in\phi} ,$$

or $\frac{\partial}{\partial \phi} \Lambda^{-1} f = i(2P-1) f + Kf$, where K is compact on each H^S . Thus we see that S is a singular integral, given by

$$S = a + ib(2P-1) + K$$

$$= (a+ib)P + (a-ib)(1-P) + K$$

Then S is Fredholm on $L^2(S^1) = H^0(S^1)$ if and only if $a + ib$ and $a - ib$ have no common zeros on S^1. However, we really want to know when S is Fredholm on H^S, especially for $S \geq 0$. The same condition suffices, as can be seen from the following proposition.

Proposition: If $a \in C^\infty(S')$, then $\Lambda^\tau a \Lambda^{-\tau} - a$ is compact on $L^2(S')$, for any real τ.

Here $\Lambda^\tau (\sum a_n e^{in\phi}) = \sum a_n (1+|n|)^\tau e^{in\phi}$. We omit the proof of this proposition here, as it is a very simple consequence of more general results proved in the next chapter. The reader might want to try to find a direct proof.*

If a and b are smooth real valued functions, then

$a \frac{\partial}{\partial r} + b \frac{\partial}{\partial \phi}$ is a vector field on S^1, and the Fredholm condition becomes the condition that this vector field not vanish at any point in S^1. In higher dimensions this oblique derivative problem becomes more subtle, and things can go wrong if the vector field is tangent to the boundary at some points. We will take a look at this in Chapter V.

* Alternatively, the Fredholm theory we worked out on $L^2(S^1)$ can easily be pushed to $H^s(S^1)$.

§3. C* algebras and singular integral operators.

The basic task of this section is to prove the necessity of the Fredholm condition given in section 1. We do this using an approach to pseudo differential operators due to Cordes; see [12], [29].

For singular intergral operators on a half line and half space, see [10], [11], and for an algebra of operators related to hypoelliptic operators see [76].

With all due apologies to the reader, we begin with the following list of algebras of operators we'll need.

$\mathcal{P}$ = norm closed algebra generated by I and P .

$\mathfrak{A}_0$ = algebra generated by P and C(S') .

$\mathfrak{A}_1$ = norm closure of $\mathfrak{A}_0$.

I = $\mathfrak{C} \cap \mathfrak{A}_1$

$\tilde{\mathfrak{A}}$ = algebra generated by $\mathfrak{A}_0$ and $\mathfrak{C}$

$\mathfrak{A}$ = norm closure of $\tilde{\mathfrak{A}}$.

Now theorem 1 of section 1 implies that the C* algebra $\mathfrak{A}_1/I$ is commutative. Hence it is isometrically isomorphic to C(M), the space of continuous complex valued functions on its maximal ideal space.

Since $\mathfrak{A}_1$ is the Banach algebra generated by $\mathcal{P}$ and $C(S^1)$, if we consider the injection $j_1 = \mathcal{P} \to \mathfrak{A}_1$ and $j_2 : C(S^1) \to \mathfrak{A}_1$, then adjoints lead to maps of maximal ideal spaces $j_1^t \times j_2^t : M \to S^1 \times \mathbb{Z}_2$

This map must be 1-1 since $\mathcal{B}$ and $C(S^1)$ generate $\mathfrak{A}_1$. It is also onto. In fact, rotational symmetry of S^1 shows that $(\phi,\varepsilon) \in M$ implies $(\phi',\varepsilon) \in M$ for any $\phi' \in S^1$, $\varepsilon = \pm 1$, and flipping S^1 about an axis in $\mathbb{R}^2$ shows that $(\phi,1) \in M \Longleftrightarrow (\phi,-1) \in M$. Thus $\mathfrak{A}_1/I \approx C(S^1 \times R_2)$.

Proposition: $\mathfrak{A}_1/I \approx \mathfrak{A}/\mathcal{C}$ in a natural manner

Proof: Consider the natural maps $\mathfrak{A}_1 \xrightarrow{j} \mathfrak{A} \longrightarrow \mathfrak{A}/\mathcal{C}$. Clearly the kernel of this composite map is precisely $\mathfrak{A}_1 \cap \mathcal{C} = I$, so we have an injective map $K: \mathfrak{A}_1/I \to \mathfrak{A}/\mathcal{C}$. Now observe that the range of $K \supset \tilde{\mathfrak{A}}/\mathcal{C}$, which is dense in $\mathfrak{A}/\mathcal{C}$ since $\tilde{\mathfrak{A}}$ is dense in $\mathfrak{A}$. But a*-homomorphism of C* algebras must have closed range (see Dixmier [16].) So K is the desired natural isomorphism.

Now the image in $C(S^1 \times \mathbb{Z}_2)$ of an element $A \in \mathfrak{A}$ under the composite map $\mathfrak{A} \to \mathfrak{A}/\mathcal{C} \approx \mathfrak{A}_1/I \approx C(S^1 \times \mathbb{Z}_2)$ is called the symbol of A and is denoted by σ_A. The reader should verify that $\sigma_a(\phi,\varepsilon) = a(\phi)$ and $\sigma_P(\phi,1) = 1$, $\sigma_P(\phi,-1) = 0$, so this agrees with the definition of σ_A given in section 1. (an immediate corollary is that every $A \in \mathfrak{A}$ can be written in the form used in section 1.)

Remark: Since an isomorphism of C* algebras preserves the norm, we have the following consequence. If $A \in \mathfrak{A}$, then

$$\inf_{K \in \mathcal{C}} ||A+K|| = \sup_{(\phi,\varepsilon) \in S^1 \times \mathbb{Z}_2} |\sigma_A(\phi,\varepsilon)| .$$

With one further result from C* algebra theory, we'll be ready to finish off the question of when is $A \in \mathfrak{A}$ Fredholm.

Proposition: If X is a C*algebra with identity I and Y is a C* subalgebra, containing I, then $A \in Y$ is invertible in Y if and only if it is invertible in X.

The reader who is familiar with the first chapter of Dixmier [16] should be able to supply the proof of this. Now for our main theorem.

Theorem: $A \in \mathfrak{A}$ is Fredholm if and only if σ_A is nowhere vanishing.

Proof: We know that A is Fredholm if and only if it is invertible in $\mathcal{L}(L^2)/\mathcal{C}$. In view of the above quoted result, this is equivalent to A being invertible in $\mathfrak{A}/\mathcal{C}$, and since $A \to \sigma_A$ is an isomorphism of $\mathfrak{A}/\mathcal{C}$ with $C(S^1 \times \mathbb{Z}_2)$, the result follows.

Exercise 5. If $T:X \to Y$ is a Fredholm operator between two Banach spaces, we define the index of T to be

$$\begin{aligned} \text{ind } T &= \dim \ker T - \dim (Y/T(X)) \\ &= \dim \ker T - \dim \ker T^* \end{aligned}$$

F(X), the set of Fredholm operators on X, is an open subset of $\mathcal{L}(X)$, with the norm topology, and the index is constant on connected components of F(X). (See [64].)

If $T = aP + b(1-P)$, where $a, b \in C(S^1)$ are non-vanishing complex valued functions, calculate ind T in terms of the winding numbers of a and b (about 0), using the above facts and explicitly calculating ind T when $a(z) = z^n$, $b(z) = z^m$ ($|z|=1$), $m, n \in \mathbb{Z}$.

Exercise 6. For $s > -\frac{1}{2}$, consider the map

$T: H^{s+2}(B) \to H^s(B) \oplus H^{s+\frac{1}{2}}(S^1)$ defined by

$u \to (\Delta u, a\frac{\partial}{\partial r} u + b\frac{\partial}{\partial \phi} u + cu \Big|_{S^1})$. If the Fredholm

condition given in section 2 is satisfied, T is Fredholm. In particular, if $a\frac{\partial}{\partial r} + b\frac{\partial}{\partial \phi}$ is a real, nonvanishing vector field, find ind T in terms of the winding number of this vector field.

Exercise 7: The range of P is a closed linear subspace of $L^2(S')$ which we will call $\underset{\sim}{H}^2$. If $\phi \in C(S^1)$, $T_\phi = P\phi|_{\underset{\sim}{H}^2}$ is an operator on $\underset{\sim}{H}^2$ called a Toeplitz operator.

(a) Show that the linear map $C(S^1) \to \mathcal{L}(\underset{\sim}{H}^2)$ given by $\phi \to T_\phi$ induces a linear * homomorphism of the algebra $C(S^1)$ into $\mathcal{L}(\underset{\sim}{H}^2)/\mathcal{C}$, where $\mathcal{C}$ in the algebra of compact operators on $\underset{\sim}{H}^2$.

(b) Show that T_ϕ is compact if and only if $\phi = 0$. Hint: T_ϕ compact $\Longrightarrow T_{|\phi|^2}$ compact. Now imitate the proof of lemma 1 in section 1.

(c) Deduce that if $\mathcal{J}$ is the C*algebra of operators on $\underset{\sim}{H}^2$ generated by $\{T_\phi : \phi \in C(S')\}$ and $\mathcal{C}$ the map $\phi \to T_\phi$ induces a *-isomorphism of $C(S^1) \to \mathcal{J}/\mathcal{C}$.

(d) Prove that $||T_\phi|| = \sup_{x \in S^1} |\phi(x)|$ for $\phi \in C(S^1)$.

Hint: Clearly $||T_\phi|| \leq ||\varphi||_\infty$; now look at $\inf_{K \in \mathcal{C}} ||T_\phi + K||$.

(e) Deduce from (c) that T_ϕ is Fredholm if and only if ϕ is nowhere vanishing, for $\phi \in C(S^1)$. Calculate the index of T_ϕ.

(f) L. Coburn [9] has shown that if $\phi \in C(S^1)$ (more generally, if $\phi \in L^\infty(S^1)$) is not identically zero, then either T_ϕ or T_ϕ^* must be injective. Using this fact and the result of exercise (e), deduce that T_ϕ is invertible if and only if ϕ is nowhere vanishing and the curve traced out by ϕ has winding number zero about the origin. More generally, describe the spectrum of T_ϕ.

CHAPTER II. PSEUDO DIFFERENTIAL OPERATORS

§1. The Fourier integral representation.

Recall that the Fourier inversion formula is

$$f(x) = (2\pi)^{-n} \int e^{i<x,\zeta>} \hat{f}(\zeta) d\zeta \ , \ f \in C_0^\infty(\mathbb{R}^n)$$

and if we differentiate this formula we get, with $D_j = \frac{1}{i} \frac{\partial}{\partial x_j}$,

$$D^\alpha f(x) = (2\pi)^{-n} \int \zeta^\alpha e^{i<x,\zeta>} \hat{f}(\zeta) d\zeta$$

Hence if $p(x,D) = \sum_{|\alpha| \leq k} a_\alpha(x) D^\alpha$ is a differential operator,

(1) $$p(x,D) f(x) = (2\pi)^{-n} \int p(x,\zeta) e^{i<x,\zeta>} \hat{f}(\zeta) d\zeta \ .$$

We shall use the Fourier integral representation (1) to define pseudo differential operators, taking the function $p(x,\zeta)$ to belong to a general class of symbols, which we now define.

Definition: Let Ω be an open subet of $\mathbb{R}^n$. Then $m, \rho, \delta \in \mathbb{R}$, $0 \leq \rho, \delta \leq 1$, $S^m_{\rho,\delta}(\Omega)$ is the set of $p \in C^\infty(\Omega \times \mathbb{R}^n)$ with the property that for any compact $K \subset \Omega$, any multi-indices α and β , there exists a constant $C_{K,\alpha,\beta}$ such that

(2) $$|D_x^\beta D_\zeta^\alpha p(x,\zeta)| \leq C_{K,\alpha,\beta}(1 + |\zeta|)^{m-\rho|\alpha| + \delta|\beta|}$$

for all $x \in K$, all $\zeta \in \mathbb{R}^n$.

If $p \in S^m_{\rho,\delta}(\Omega)$ we say p is a symbol on Ω of order m, and type (ρ,δ). If p satisfies (2) only for $|\zeta| \geq R$, we say $p \in S^m_{\rho,\delta}$ for large ζ.

Exercise 1. If $p(x,D)$ is a differential operator of order m, show that $p(x,\zeta) \in S^m_{1,0}$.

If $p(x,\zeta) \in C^\infty(\Omega \times (\mathbb{R}^n \setminus 0))$ is homogeneous of degree m in ζ, prove that $p \in S^m_{1,0}$ for large ζ.

Exercise 2. $p \in S^m_{\rho,\delta}(\Omega) \Longrightarrow p^{(\alpha)}_{(\beta)} = i^{|\beta|} D^\beta_x D^\alpha_\zeta p \in S^{m-\rho|\alpha|+\delta|\beta|}_{\rho,\delta}(\Omega)$.

If $q \in S^{m_1}_{\rho,\delta}(\Omega)$, then $p(x,\zeta)\, q(x,\zeta) \in S^{m+m_1}_{\rho,\delta}(\Omega)$.

If $p \in S^0_{\rho,\delta}(\Omega)$ and ϕ is smooth in a nbd of the closure of the set of values assumed by p, then $\phi(p(x,\zeta)) \in S^0_{\rho,\delta}(\Omega)$.

Theorem: If $p \in S^m_{\rho,\delta}(\Omega)$ then $p(x,D)$ defined by formula (1) is a continuous map of $C^\infty_0(\Omega)$ into $C^\infty(\Omega)$. If $\delta < 1$, then the map can be extended to a continuous map

$$p(x,D): \mathcal{E}'(\Omega) \to \mathcal{D}'(\Omega) .$$

Proof: If $p \in S^m_{\rho,\delta}(\Omega)$, $u \in C^\infty_0(\Omega)$, then the integral

$$p(x,D)u = (2\pi)^{-1} \int p(x,\zeta) e^{i\langle x,\zeta\rangle} \hat{u}(\zeta)\, d\zeta$$

is absolutely convergent, and it is okay to differentiate under the integral sign. That $p(x,\eta) = C^\infty_0(\Omega) \to C^\infty(\Omega)$ follows.
In order to prove the rest of the theore m, we need a lemma.

Lemma: Let $p \in S^m_{\rho,\delta}(\Omega)$, $v \in C^\infty_0(\Omega)$. Then $\forall\, \zeta, \eta \in \mathbb{R}^n$,

$$\left|\int v(x)\, p(x,\xi) e^{i<x,\eta>}\, dx\right| \leq C_N (1+|\xi|)^{m+\delta N} (1+|\eta|)^{-N} .$$

Proof of lemma: Integrating by parts yields

$$\left|\eta^\alpha \int v(x)\, p(x,\xi) e^{i<x,\,\eta>}\, dx\right| = \left|\int D_x^\alpha (v(x)p(x,\xi)) e^{i<x,\,\eta>} dx\right|$$

$$\leq C_{|\alpha|} (1+|\xi|)^{m+\delta|\alpha|}$$

since $|D_x^\beta p(x,\zeta)| \leq C(1+|\zeta|)^{m+\delta|\beta|}$.

To complete the proof of the theorem, we show that the functional $v \longrightarrow <p(x,D)u,v>$ is well defined if $u \in \mathcal{E}'$.

(3) $$<p(x,D)u,v> = (2\pi)^{-n} \int p(x,\zeta)\, \hat{u}(\zeta) e^{i<x,\zeta>}\, v(x)\, d\zeta\, dx$$

$$= (2\pi)^{-n} \int p_v(\zeta)\, \hat{u}(\zeta)\, d\zeta$$

where $p_v(\zeta) = \int v(x)\, p(x,\zeta)\, e^{i<x,\zeta>}\, dx$. For (3) to make sense for $u \in \mathcal{E}'(\Omega)$, we only need that $p_v(\zeta)$ be rapidly decreasing. But by the lemma, with $\zeta = \eta$, we have

$$|p_v(\zeta)| \leq C_N (1+|\zeta|)^{m+(\delta-1)N} .$$

Since $\delta < 1$, we see that $p_v(\zeta)$ is rapidly decreasing, as desired, so $p(x,D)u \in \mathcal{D}'(\Omega)$ is defined for $u \in \mathcal{E}'(\Omega)$.

Definition: If $p \in S^m_{\rho,\delta}(\Omega)$, then the operator $p(x,D)$ defined by (1) is called a pseudo differential operator, and we write $p(x,D) \in PS(m,\rho,\delta)$.

In the remaining sections of this chapter we examine some of the fundamental properties of pseudo differential operators. Our treatment follows closely that of Hormander [34] and [39], where many other results are proved, and we urge the reader to look at these papers.

§2. The pseudo local property.

If $K \in \mathcal{D}'(\Omega x \Omega)$, then the map $K: C_0^\infty(\Omega) \longrightarrow \mathcal{D}'(\Omega)$ is given by $\langle Ku, v\rangle = \langle K, u(x)v(y)\rangle$.

Theorem on Singular Support: Suppose $K \in \mathcal{D}'(\Omega x \Omega)$ satisfies

(i) $K: C_0^\infty(\Omega) \longrightarrow C^\infty(\Omega)$

(ii) $K: \mathcal{E}'(\Omega) \to \mathcal{D}'(\Omega)$

(iii) K is C^∞ off the diagonal in $\Omega \times \Omega$.

If $u \in \mathcal{E}'(\Omega)$ is smooth on $\omega \subset \Omega$, then Ku is smooth on ω, or sing supp $Ku \subset$ sing supp. u .

The proof of this result we leave to the reader as as exercise in distribution theory. We apply it to the case of pseudo differential operators.

Theorem: If $p(x,\zeta) \in S^m_{\rho,\delta}(\Omega)$, $u \in \mathcal{E}'(\Omega)$, Then if $\rho > 0$ we have

(1) sing supp $p(x,D)u \subset$ sing supp u .

Proof: We show that the distribution kernel of $p(x,D)$ is smooth off the diagonal in $\Omega \times \Omega$. Call the Kernel K .

$$K(u \otimes v) = \langle p(x,D)u, v\rangle$$
$$= \int v(x)p(x,D)u(x)dx$$
$$= \iint p(x,\zeta)e^{i\langle x,\zeta\rangle} v(x)\,\hat{u}(\zeta)\,dx\,d\zeta$$

For any $\omega \in C_0^\infty(\Omega x \Omega)$, with $\hat{\omega}(x,\zeta) = \int \omega(x,y)e^{-i\langle y,\zeta\rangle}dy$, we get

$$<K,w> = \iint \hat{w}(x,\xi)\, p(x,\xi)\, e^{i<x,\xi>}\, dx\, d\xi \; .$$

$$<(x-y)^{\alpha}K,w> = \iint \left[(x-D_{\xi})^{\alpha}\, \hat{w}(x,\xi)\right] p(x,\xi) e^{i<x,\xi>} dx\, d\xi$$

$$= \iint \hat{w}(x,\xi) \left[(x-D_{\xi})^{\alpha}\; e^{i<x,\xi>}\, p(x,\xi) \right] dx\, d\xi$$

$$= \iint \hat{w}(x,\xi) e^{i<x,\xi>}\; D_{\xi}^{\alpha}\, p(x,\xi)\, dx d\xi$$

using Leibniz' formula. Hence

$$<(x-y)^{\alpha}\; K,w> \; = \iint w(x,y) e^{i<x-y,\xi>}\; D_{\xi}^{\alpha}\, p(x,\xi)\; dx\; d\xi\; dy \; .$$

If α is so large that $m - \rho|\alpha| < -n -j$, it follows that $(x-y)^{\alpha}\; K \in C^{j}(\Omega x \Omega)$, since

$$(x-y)^{\alpha} K_{x,y} = \int e^{i<x-y,\xi>}\; D_{\xi}^{\alpha} p(x,\xi)\; d\xi \, ,$$

and this integral converges for α large enoough Hence K is smooth off the diagonal and we are done.

Corollary: If $p \in S^{m}_{\rho,\delta}(\Omega)$, $n = \dim \Omega$, $\delta < 1$, and $m + n + j < 0.$, then $K \in C^{j}(\Omega x \Omega)$.

Proof: Take $\alpha = 0$ in the proof of the above theorem.

The property (1) we proved in the above theorem is frequently called the pseudo local property of pseudo differential operators. An important consequence of the Corollary proved above is that an operator $T \in PS(-\infty,\rho,\delta) = \bigcap_{m>-\infty} PS(m,\rho,\delta)$ is a smoothing operator, since its distribution kernel is smooth.

§3. Asymptotic expansions of a symbol.

Theorem 1: Suppose $p_j \in S^{m_j}_{\rho,\delta}(\Omega)$, $m_j \downarrow -\infty$. Then there exists $p \in S^{m_0}_{\rho,\delta}(\Omega)$ such that $p - \sum_{j \geq m_k} p_j \in S^{m_k}_{\rho,\delta}(\Omega)$.

In this case we write $p \sim \sum p_j$.

Proof: Pick compact sets K_j with $K_1 \subset K_2 \subset \cdots \to \Omega$. Pick $\phi \in C^\infty(\mathbb{R}^n)$ with $\phi(\xi) = 0$ for $|\xi| \leq \frac{1}{2}$, $\phi(\xi) = 1$ for $|\xi| \geq 1$. Our desired function will have the form

$$p(x,\xi) = \sum_{j=1}^{\infty} \phi\left(\frac{\xi}{t_j}\right) p_j(x,\xi)$$

where the t_j are picked so large that $\left|D_x^\beta D_\xi^\alpha \phi\left(\frac{\xi}{t_j}\right) p_j(x,\xi)\right| \leq 2^{-j}(1 + |\xi|)^{m_j+1-\rho|\alpha|+\delta|\rho|}$ for $|\alpha| + |\beta| + i \leq j$ and $x \in K_i$. We leave the details to the reader.

Exercise 3. Let $\sum_{j=0}^{\infty} a_j x^j$ be an arbitrary formal power series. Show that there exists an $f \in C^\infty(\mathbb{R})$ such that $f^{(j)}(0) = j!a_j$, hence $f(x) - \sum_{j=1}^{k} a_j x^j = 0(x^{k+1})$

This classical result, due to Borel, can be proved in a similar to the proof of the above theorem.

The next theorem shows that the asymptotic relation $p \sim \sum p_j$ is valid if a weaker asymptotic relation is assumed to hold. It also gives a useful method of proving that a function $p \in C^\infty(\Omega \times \mathbb{R}^n)$ actually is a symbol, in many cases.

Theorem 2: Let $p_j \in S^{m_j}_{\rho,\delta}(\Omega)$, $m_j \downarrow -\infty$. Let $p \in C^\infty(\Omega \times \mathbb{R}^n)$ and assume that for all multi-indices α,β there exists μ such that

$$|p^{(\alpha)}_{(\beta)}(x,\xi)| \leq C_k(1+|\xi|)^\mu \qquad x \in K \subset\subset \Omega .$$

If there exist $\mu_k \to -\infty$ such that

$$|p(x,\xi) - \sum_{j<k} p_j(x,\xi)| \leq C_K(1+|\xi|)^{-\mu_k}, \ x \in K$$

then $p \in S^{m_0}_{\rho,\delta}(\Omega)$, $p \sim \sum p_j$.

Proof: By theorem 1 we can find $q \in S^{m_0}_{\rho,\delta}(\Omega)$ such that $q \sim \sum p_j$. It only remains to show that $p-q \in S^{-\infty}_{\rho,\delta}$.

From the hypotheses we immediately see that, for each μ,

$$|p(x,\xi)| \leq C_{K,\mu}(1+|\xi|)^{-\mu}, \ x \in K.$$

We need only verify that such an inequality holds for $p^{(\alpha)}_{(\beta)} - q^{(\alpha)}_{(\beta)}$. For this, we use the inequality

$$(*) \qquad \sum_{|\alpha|=1} \sup_{K_1} |D^\alpha f|^2 \leq C \sup_{K_2}|f| \sum_{|\alpha|\leq 2} \sup_{K_2}|D^\alpha f| ,$$

where $K_1 \subset \operatorname{int} K_2 \subset K_2$, K_j compact

If we apply this inequality to the functions $(x,\eta) \mapsto p(x,\xi+\eta) - q(x,\xi+\eta)$, considering ξ as a parameter, and taking $K_1 = K \times \{0\}$ K_2 small nbd of K_1, we get

$$\sup_{x \in K} |D_\nu p(x,\xi) - D_\nu q(x,\xi)|$$

$$\leq C \sup_{(x,\eta) \in K_2} |p(x,\xi+\eta) - q(x,\xi+\eta)| \left(\sum_{j \leq 2} \sup_{(x,\eta) \in K_2} \left| D_\nu^j p(x,\xi+\eta) - D_\nu^j q(x,\xi+\eta) \right| \right) \leq C'_\mu (1 + |\xi|)^{-\mu}$$

since the first factor is rapidly decreasing and the second factor has polynomial growth. Here D_ν denotes differentiation in any given x or ξ direction. Inductively it follows that such an inequality is valid for $p_{(\beta)}^{(\alpha)} - q_{(\beta)}^{(\alpha)}$ with any multi-indices α and β, and the proof is complete.

Before we continue with results on pseudo differential operators, let us say a word about the inequality (*) .

Proposition: Let the closed linear operator A generate a contraction semigroup on a Banach space X. Then for $u \in \mathcal{D}(A^2)$ we have

$$||Au||^2 \leq 4||u|| \, ||A^2u||$$

Proof: From the identity

$$-t\,Au = t(t-A)^{-1} A^2u + t^2u - t^2 t(t-A)^{-1}u$$ and

the inequality $||t(t-A)^{-1}|| \leq 1$, valid for the generator of a contraction semigroup, we get

$$t\,\|Au\| \leq \|A^2u\| + 2t^2\,\|u\|\,, \text{ or}$$

$$\|Au\| \leq \inf_{t\geq 0} \left(\frac{1}{t}\,\|A^2u\| + 2t\,\|u\|\right)$$

$$= 2\,\|A^2u\|^{1/2}\,\|u\|^{1/2}.$$

Corollary: $\|D_j u\|_{L^\infty}{}^2 \leq 4\,\|u\|_{L^\infty}\,\|D_j{}^2 u\|_{L^\infty}$ for all $u \in C_0^\infty(\mathbb{R}^n)$.

Proof: Apply the above proposition to the Banach space $C_0(\mathbb{R}^n)$ and the obvious translation group.

Exercise 4. Prove the inequality (*).

Our final goal in this section is to introduce a class of operators, apparantly more general than the class of pseudo differential operators defined in Section 1 and then prove that there has been no essential generalization. This will make for greater freedom in constructing pseudo differential operators which will be very useful in succeeding sections.

The operators we consider will be of the form

(1) $\quad A\,u(x) = (2\pi)^{-n} \int e^{i\langle x-y,\xi\rangle}\, a(x,y,\xi)\, u(y)\, dy\, d\xi,$

$a \in S^m_{\rho,\delta}\,(\Omega\times\Omega,\ \mathbb{R}^n)$. The definition of this new symbol space is very much like the old; we require

$$|D^\beta_{x,y}\ D^\alpha_\xi\ a(x,y,\xi)| \leq C_K(1 + |\xi|\,)^{m-\rho|\alpha| + \delta|\beta|},\ (x,y) \in K.$$

It is easy to see that $A\colon C_0^\infty(\Omega) \to C^\infty(\Omega)$.

Def: $\sum$ is a proper subset of $\Omega \times \Omega$ if $\{(x,y) \in \sum\colon x \in K \text{ or } y \in K\}$ is compact whenever K is compact.

Def: A distribution $A \in \mathcal{E}'(\Omega \times \Omega)$ is said to be properly supported if suppA is a proper subset of $\Omega \times \Omega$.

It is easy to see that if $A\colon C_0^\infty(\Omega) \to C^\infty(\Omega)$ is properly supported, then we can extend A to a map $A\colon C^\infty(\Omega) \to C^\infty(\Omega)$.

Theorem 3: Let A be defined by (1). If A is properly supported, and if $0 \le \delta < \rho \le 1$, then there is $p(x,\xi) \in S^m_{\rho,\delta}(\Omega)$ such that $A\,u(x) = p(x,D)\,u$. In fact $p(x,\xi) = e^{-\langle x,\xi\rangle}A(e^{i\langle x,\xi\rangle})$, and

$$p \sim \sum_{\alpha \ge 0} \frac{1}{\alpha!}(iD_\xi)^\alpha D_y^\alpha\, a(x,y,\xi)\big|_{y=x}.$$

Proof: By the previous remark, $p(x,\xi) = e^{-\langle x,\xi\rangle}A(e^{i\langle x,\xi\rangle})$ is a well defined smooth function of its arguments, and if we apply the linear operator A to

$$u(x) = (2\pi)^{-n}\int \hat{u}(\xi)\, e^{i\langle x,\xi\rangle}d\xi, \text{ we get}$$

$$A\,u(x) = (2\pi)^{-n}\int \hat{u}(\xi)\, e^{i\langle x,\xi\rangle}p(x,\xi)\,d\xi.$$

It remains only to show that $p \in S^m_{\rho,\delta}(\Omega)$.

Note that if $b(x,y,\eta) = a(x, x+y,\eta)$ and $\hat{b}(x,\xi,\eta) = \int b(x,y,\eta)\, e^{-i<y,\xi>} dy$, then $p(x,\eta) = \int \hat{b}(x,\xi,\eta+\xi)\, d\xi$.

But $|D^{\alpha,\beta}_{x,y} D^{\gamma}_{\eta}\, b(x,y,\eta)| \le C_K(1 + |\eta|)^{m+\delta(|\alpha| + |\beta|) - \rho|\gamma|} (x \in K)$

Since $a \in S^m_{\rho,\delta}$ and is properly supported. Hence for any integer ν we have

$$(2) \quad |D^{\alpha}_{x} D^{\gamma}_{\eta}\ \hat{b}(x,\xi,\eta)| \le C_{K,\alpha,\gamma,\nu}(1 + |\eta|)^{m+\delta(|\alpha| + \nu) - \rho|\gamma|} (1 + |\xi|)^{-\nu}.$$

Since $\delta < 1$ it follows that $p(x,\eta)$ and any of its derivatives can be bounded by some power of $1 + |\eta|$. We now set things up to apply Theorem 2.

If we take the Taylor expansion of $\hat{b}(x,\xi,\eta+\xi)$ in the last argument, about the point η, (2) yields

$$|\hat{b}(x,\xi,\eta+\xi) - \sum_{|\alpha|<N} (iD_{\eta})^{\alpha}\ \hat{b}(x,\xi,\eta) \frac{1}{\alpha!} \xi^{\alpha}|$$

$$\le C_{\nu}|\xi|^{N} \sup_{0 \le t \le 1} (1 + |\eta + t\xi|)^{m + \delta\nu - \rho N}$$

$$(1 + |\xi|)^{-\nu}$$

Here ν can be 0 or any positive integer. With $\nu = N$ we obtain a bound

$$C(1 + |\eta|)^{m+(\delta-\rho)N} \quad \text{if} \quad |\xi| < \tfrac{1}{2}|\eta|,$$

and if ν is large we get a bound by any power of $(1 + |\xi|)^{-1}$ for $|\eta| < 2|\xi|$. Hence

$$\left| p(x,\eta) - \sum_{|\alpha| \leq N} (iD_\eta)^\alpha D_y^\alpha b(x,y,\eta) \frac{1}{\alpha!} \Big|_{y=x} \right|$$
$$\leq C(1 + |\eta|)^{m+n+(\delta-\rho)N} .$$

Theorem 3 now follows immediately from theorem 2.

Properly supported operators can be composed, and the composition is properly supported. Since the distribution kernel of any pseudo differential operator is smooth off the diagonal, we may write such an operator as the sum of a property supported operator and a smoothing operator. In the future, pseudo differential operators will usually be assumed to be properly supported, often without comment.

4. Adjoints and products.

Theorem 1: If $p(x,D) \in PS(m,\rho,\delta)$ is properly supported and if $0 \leq \delta < \rho \leq 1$, then $p(x,D)^t \in PS(m,\rho,\delta)$ and $p(x,D)^t = q(x,D)$ with

$$q(x,\xi) \sim \sum_{\alpha} \frac{1}{\alpha!} (iD_\xi)^\alpha D_x^\alpha p(x, -\xi)^t.$$

Proof: If $A = p(x,D)$, we have

$$\begin{aligned} \langle Au, v \rangle &= (2\pi)^{-n} \int v(y) \int e^{i\langle y-x,\xi\rangle} p(y,\xi) u(x)\, dx\, d\xi\, dy \\ &= (2\pi)^{-n} \int u(x) \int e^{i\langle y-x,\xi\rangle} p(y,\xi)^t v(y)\, dy d\xi dx \end{aligned}$$

Hence $A^t v(x) = (2\pi)^{-n} \int e^{i\langle y-x,\xi\rangle} p(y,\xi)^t v(y)\, dy d\xi$

(1) $\qquad = (2\pi)^{-n} \int e^{i\langle x-y,\xi\rangle} p(y,-\xi)^t v(y)\, dy d\xi.$

If we apply Theorem 3 of the previous section, with $a(x,y,\xi) = p(y,-\xi)$, the proof is easily completed.

If A in a pseudo differential operator, we will often write $A = \sigma_A(x,D)$.

Theorem 2: If $A \in PS(m,\rho,\delta)$ and $B \in PS(m',\rho,\delta)$ are properly supported and if $0 \leq \delta < \rho \leq 1$, then $BA \in PS(m+m',\rho,\delta)$ and

$$\sigma_{BA}(x,\xi) \sim \sum_{\alpha \geq 0} \frac{1}{\alpha!} [(iD_\xi)^\alpha \sigma_B(x,\xi)]\, D_x^\alpha \sigma_A(x,\xi). \tag{2}$$

Proof: Applying (1) to the operator A^t, we have

$$Au(x) = A^{tt} u(x) = (2\pi)^{-n} \int e^{i\langle x-y,\xi\rangle} \sigma_{A^t}(y,-\xi)^t u(y)\, dy d\xi,$$

or $\hat{Au}(\xi) = \int e^{-i\langle y,\xi\rangle} \sigma_{A^t}(y,-\xi)^t u(y)\, dy.$

$$\therefore BA\, u(x) = (2\pi)^{-n} \int e^{i\langle x,\xi\rangle} \sigma_B(x,\xi)\, \hat{Au}(\xi)\, d\xi$$

$$= (2\pi)^{-n} \int e^{i\langle x-y,\xi\rangle} \sigma_B(x,\xi)\, \sigma_{A^t}(y,-\xi)^t u(y)\, dyd\xi.$$

Applying theorem 3 of the last section, with $a(x,y,\xi) = \sigma_B(x,\xi)\, \sigma_{A^t}(y,-\xi)^t$, we get $BA \in PS(m+m',\rho,\delta)$ with

$$\sigma_{BA}(x,\xi) \sim \sum_{\alpha\geq 0} \frac{1}{\alpha!} (iD_\xi)^\alpha D_y^\alpha \sigma_B(x,\xi)\, \sigma_{A^t}(y,-\xi)^t\big|_{y=x}$$

and a simple manipulation puts this in the form (2).

We note as a consequence of these two theorems the following two facts:

(i) $\sigma_{A^t}(x,\xi) - \sigma_A(x,\xi)^t \in S^{m-(\rho-\delta)}_{\rho,\delta}$

(ii) $\sigma_{BA}(x,\xi) - \sigma_B(x,\xi)\, \sigma_A(x,\xi) \in S^{m+m'-(\rho-\delta)}_{\rho,\delta}$.

From (ii) it follows that $BA - AB \in PS(m+m'-(\rho-\delta),\rho,\delta)$, provided the symbols σ_A and σ_B are scalar, or commuting matrices.

Exercise: If $A \in PS(m,\rho,\delta)$, $0\leq\delta<\rho\leq 1$, and if $\varphi,\psi \in C_c^\infty(\Omega)$, with disjoint supports, show that

$$\psi A \varphi \in PS(-\infty,\rho,\delta).$$

Use this to rederive the pseudo local property of A.

5. Coordinate changes, operators on a manifold.

We now shall see what happens to a pseudo differential operator under a change of coordinates. Let Ω and X be regions in $\mathbb{R}^n$, and let $\chi:\Omega \to X$ be a diffeomorphism. If $A = p(x,D) \in PS(m,\rho,\delta;\Omega)$, then $A:C_0^\infty(\Omega) \to C^\infty(\Omega)$, so if we set $A_1 u = A(u \circ \chi)\circ\chi^{-1}$, $A_1:C_0^\infty(X) \to C^\infty(X)$.

Theorem: If $1-\rho \leq \delta < \rho$, we have $A_1 \in PS(m,\rho,\delta;X)$.

(†) Furthermore, $\sigma_{A_1}(\chi(x),\xi) \sim \sum_{\alpha\geq 0} \frac{1}{\alpha!}\phi_\alpha(x,\xi)\, p^{(\alpha)}(x, {}^t\chi'(x)\xi)$

where $\phi_\alpha(x,\xi) = D_y^\alpha \exp(i<\chi(y) - \chi(x) - \chi'(x)(y-x),\xi>)|_{x=y}$

is a polynomial in ξ of degree $\leq \frac{1}{2}|\alpha|$.

Proof: $A_1 u(x) = (2\pi)^{-n} \int e^{i<\chi_1(x) - y,\xi>} \sigma_A(\chi_1(x),\xi)\, u(\chi(y))\, dyd\xi$

$$= (2\pi)^{-n} \int e^{i<\chi_1(x)-\chi_1(y),\xi>} \sigma_A(\chi_1(x),\xi)\, u(y) \left(\frac{\partial\chi_1}{\partial y}\right) dyd\xi$$

$$= (2\pi)^{-n} \int e^{i<(x-y),{}^t\Phi(x,y)\xi>} \sigma_A(\chi_1(x),\xi) \left(\frac{\partial\chi_1}{\partial y}\right) u(y)\, dyd\xi.$$

Here we have set $\chi_1 = \chi^{-1}$, and Φ is defined by setting

$$<\chi_1(x) - \chi_1(y),\xi> = \Sigma(\chi_1^{\,j}(x) - \chi_1^{\,j}(y))\,\xi_j$$

$$= \Sigma\Sigma\, \Phi_{kj}(x,y)\,(x_k - y_k)\,\xi_j.$$

Then Φ is defined and smooth near the diagonal in $X \times X$.

(*) Thus $A_1\ u(x) = (2\pi)^{-n} \int e^{i<x-y,\xi>} \sigma_A(\chi_1(x), \psi(x,y)\xi)$ $D(x,y)\ u(y)\ dyd\xi + K$. Here we have set $\psi(x,y) = \phi^{-1}(x,y)$ and $D(x,y) = \det(\chi_1{}') \det(\psi(x,y))\ \Xi(x,y)$ where $\Xi(x,y)$ is identically 1 in a neighborhood of the diagonal in X, and is thrown in because $\psi(x,y)$ might not be defined everywhere. The resulting error term K is a smoothing operator.

Now in formula (*) we have one familiar representation for a pseudo differential operator, with

$$a(x,y,\xi) = \sigma_A(\chi_1(x),\ \psi(x,y)\xi)\ D(x,y). \tag{1}$$

Hence A_1 is a pseudo differential operator with symbol

$$\sigma_{A_1}(x,\xi) \sim \sum_{\alpha\geq 0} \frac{1}{\alpha!} (iD_\xi)^\alpha\ D_y^\alpha\, \sigma_A(\chi_1(x),\psi(x,y)\xi)\ D(x,y)\big|_{y=x}. \tag{2}$$

Exercise: Prove that $a(x,y,\xi)$ defined by (1) does belong to $S^m_{\rho,\delta}(\Omega\times\Omega;\ \mathbb{R}^n)$. You'll need the hypothesis $1-\rho \leq \delta < \rho$.

We can write this last sum in the form

$$\sigma_{A_1}(x,\xi) \sim \sum_{\gamma,\beta} c_\gamma(x,y)\xi^\gamma\ \sigma_A^{(\beta)}\ (\chi_1(x),\psi(x,y)\xi)\big|_{y=x}. \tag{3}$$

But if one inspects (2) for the places where one can obtain factors of ξ^γ, one finds that in the sum (3), $|\gamma| < \frac{1}{2}|\beta|$.

(4) Hence $\sigma_{A_1}(\chi(x),\xi) \sim \sum_{\alpha\geq 0} \frac{1}{\alpha!}\ \phi_\alpha(x,\xi)\ p^{(\alpha)}(x,{}^t\chi'(x)\xi)$

with $\phi_\alpha(x,\xi)$ polynomials in ξ of degree $\leq \frac{1}{2}|\alpha|$, since $\psi(x,x) = {}^t\chi'(x)$. That the term $\phi_\alpha(x,\xi)$ have the indicated form can be checked by taking A to be specified operator, since these functions are independent of the operator A. Taking $A = I$ we see that $\phi_0(x,\xi) = 1$. Also, $\phi_\alpha = 0$ if $|\alpha| = 1$. We won't need any higher term, so we refer the reader to [34].

Note that the general term in (4) belongs to $S_{\rho,\delta}^{m-(\rho-\frac{1}{2})|\alpha|}$, so to say that this series is asymptotic, we need $\rho > \frac{1}{2}$, which follows from the inequality $1 - \rho \leq \delta < \rho$.

We now define the concept of a pseudo differential operator on a manifold X. Namely $A: C_0^\infty(X) \to C^\infty(X)$ belongs to $PS(m,\rho,\delta; X)$ if the distribution kernel of A is a smooth off the diagonal and if for any coordinate neighborhood U in X, with $\chi: U \to V$ a diffeomorphism onto an open subset V of $\mathbb{R}^n$, the resulting map of $C_0^\infty(V) \to C^\infty(V)$ given by $u \mapsto A(u \circ \chi) \circ \chi^{-1}$ belongs to $PS(m,\rho,\delta; V)$. We assume that $1 - \rho \leq \delta < \rho$.

If $p(x,D) \in PS(m,\rho,\delta; V)$ and V is an open subset of $\mathbb{R}^n$, we define the principal symbol of $P(x,D)$ to be the equivalence class of $p(x,\xi)$ in $S_{\rho,\delta}^m(V)/S_{\rho,\delta}^{m-2(\rho-\frac{1}{2})}(V)$. We shall also call any member of this equivalence class (for large ξ) a principal symbol of $p(x,D)$.

Since the leading term in the expansion (†) for $\sigma_{A_1}(\chi(x),\xi)$ is $p(x, {}^t\chi'(x)\xi)$, we conclude that if A is a pseudo differential operator on a manifold X, its principal symbol is a well defined function on the cotangent bundle $T^*(X)$.

More generally, we could say an operator $A: C_0^\infty(X) \to C^\infty(X)$ belongs to $\overline{PS}(m,\rho,\delta;X)$ if it can be written as the sum of a smoothing operator and an operator defined as above. (*)

Exercise 5: Let $T: C_0^\infty(\mathbb{R}) \to C^\infty(\mathbb{R})$ be defined by

$$T\,u(x) = \frac{1}{\pi i}\ PV \int_{-\infty}^{\infty} \frac{u(y)}{x-y}\,dy.$$

$$= \frac{1}{\pi i}\, u * PV(\tfrac{1}{x}).$$

Taking the Fourier transform of $PV(\frac{1}{x})$, show that $T \in \overline{PS}(0,1,0)$ and find its principal symbol.

Exercise 6: Recall the operator P in $L^2(S^1)$ defined in Chapter 1 by $P\left(\sum_{n=-\infty}^{\infty} a_n\, e^{in\phi}\right) = \sum_{n=0}^{\infty} a_n\, e^{in\phi}$.

Prove that $P \in \overline{PS}(0,1,0;S^1)$, and find its principal symbol. (Hint: the operator T defined in exercise 5 bears a close relationship to the Hilbert transform H defined in Chapter 1. Use the Cayley transform of the upper half plane onto the disc and analytic properties of the operator T and P to pass from exercise 5 to exercise 6.)

Exercise 7: Prove that a properly supported smoothing operator belongs to $PS(-\infty,1,0)$.

(*) We shall gloss over this distinction in the text, hopefully without discomfort to the reader.

6. Continuity on H^S.

The basic aim of this section is to prove that if $A \in PS(m,\rho,\delta)$ and $\delta < \rho$ then $A : H^S_{comp} \to H^{S-m}_{loc}$. Since the proof for $\delta = 0$ is more straightforward than the proof in the general case, we give that first.

Theorem 1: If $p(x,D) \in PS(0,\rho,0)$ is properly supported then for any given compact set K we have the inequality

$$\int_K |p(x,D)\, u|^2\, dx \le C_K \, \|u\|^2_{L^2} \quad \forall u \in C_0^\infty(K).$$

Proof: We proved in section 1 of this chapter that

$$\left|\int p_f(x,\xi)\, e^{i\langle x,\eta\rangle}\, dx\right| \le C_N (1 + |\xi|)^{\delta N}\, (1 + |\eta|)^{-N},$$

where $p_f(x,\xi) = f(x)\, p(x,\xi)$, $f \in C_0^\infty(\Omega)$. Take $f \equiv 1$ on K. Thus for $u,v \in C_0^\infty(K)$,

$$\begin{aligned} |\langle p(x,D)\, u,v\rangle| &= \left|\iint v(x)\, p(x,\xi)\, e^{i\langle x,\xi\rangle}\, \hat{u}(\xi)\, d\xi dx\right| \\ &= \left|\iint \hat{v}(-\eta)\, \hat{p}_f(\eta - \xi,\xi)\hat{u}(\xi)\, d\xi d\eta\right| \\ &\le C_N \iint |\hat{v}(-y)\, \hat{u}(\xi)|\, (1 + |y - \xi|)^{-N}\, d\xi dy \\ &\le C'\, \|v\|_{L^2}\, \|u\|_{L^2} \quad \text{if } N \text{ is taken large.} \end{aligned}$$

This completes the proof. Note that it breaks down for nonzero δ.

The case of nonzero δ is somewhat more subtle. The following ingenious proof is due to Hörmander [39]. Even though we will make no use of this more refined continuity result in this course, we shall give the proof here, first because it is a very beautiful variation on the idea that a positive linear functional λ on C(K), the set of continuous function on a compact Hausdorff space K, is automatically continuous, with $\|\lambda\| = \lambda(1)$, and second because a good bit of the work we do here would have to be done anyway, and we shall see the following lemma revealed in section 8 as Gårding's inequality.

Lemma: If $p \in S^{0}_{\rho,\delta}(\Omega)$, $\delta < \rho$, and if $\operatorname{Re} p(x,\xi) \geq C > 0$, then there exists a $B \in PS(0,\rho,\delta)$ such that

$$\operatorname{Re} p(x,D) - B^* B \in PS(-\infty,\rho,\delta).$$

Proof: We write the symbol $b(x,\xi) \sim \sum b_j(x,\xi)$ as an asymptotic series, with $b_j \in S^{-j(\rho-\delta)}_{\rho,\delta}$. We start with

$$b_0(x,\xi) = \sqrt{\operatorname{Re} p(x,\xi)}.$$

By exercise 2 of this chapter, $b_0 \in S^{0}_{\rho,\delta}(\Omega)$. Furthermore, the formulas for adjoints and products show that

$$\operatorname{Re} p(x,D) - b_0(x,D)^* b_0(x,D) \in PS(-(\rho-\delta),\rho,\delta).$$

Proceeding by induction, suppose we have the terms $b_0,\dots,b_j$ in the asymptotic expansion. We need $b_{j+1} \in S^{-(j+1)(\rho-\delta)}_{\rho,\delta}(\Omega)$ such that

$$(*) \qquad \operatorname{Re} p(x,D) = \Big((b_0^* + \dots + b_j^*) + b_{j+1}^*\Big)\Big((b_0 + \dots + b_j) + b_{j+1}\Big) + R_{j+1}$$

with $R_{j+1} \in PS(-(j+1)(\rho-\delta),\rho,\delta)$. The right hand side of the above expression is equal to

$$R_j + b_{j+1}^* (b_0+\ldots+ b_j+b_{j+1}) + (b_0^*+\ldots+b_{j+1}^*) b_{j+1}$$

$$= R_j + b_{j+1}^* b_0 + b_0^* b_{j+1} + S_j$$

where $R_j \in PS(-j(\rho-\delta),\rho,\delta)$ is the analogous remainder term at the previous stage of (*), and $S_j \in PS(-(j-1)(\rho-\delta),\rho,\delta)$. Then it only remains to set $b_{j+1} = -\frac{1}{2} b_0^{-1} R_j$, and the induction is complete. It is easy to verify that the resulting operator $B = b(x,D)$ has the desired property.

Theorem 2: Let $A \in PS(0,\rho,\delta)$ and assume that $\delta < \rho$. Suppose $\limsup_{|\xi|\to\infty} |\sigma_A(x,\xi)| < M < \infty$. If $K \subset\subset \Omega$, there is an $R \in PS(-\infty,\rho,\delta)$ such that, for all $u \in C_0(K)$,

$$(Au,Au) \le M^2 \|u\|^2 + (Ru,u)$$

Proof: $C = M^2 - A^*A$ has principal symbol $\sigma_C = M^2 - |\sigma_A(x,\xi)|^2$, so by the previous lemma there is a $B \in PS(0,\rho,\delta)$ such that

$$C - B^* B = M^2 - A^* A - B^* B = R_0 \in PS(-\infty,\rho,\delta).$$

$$\therefore (Au,Au) \le (Au,Au) + (Bu,Bu)$$

$$= M^2 \|u\|^2 + (R_0u,u),$$

and the proof is complete.

Corollary: If $A \in PS(0,\rho,\delta)$ is properly supported, then for $K \subset\subset \Omega$, $A: L^2(K) \to L^2_{loc}(\Omega)$ cont. Hence $A: L^2_{loc}(\Omega) \to L^2_{loc}(\Omega)$ continuously. Furthermore, if $\lim_{|\xi|\to\infty} \sigma_A(x,\xi) = 0$, then $A = L^2(K) \to L^2_{loc}(\Omega)$ is compact.

Exercise 8: Prove the proposition stated at the end of section 2 in chapter 1.

Exercise 9: Define Λ^s on $\mathcal{S}'(\mathbb{R}^n)$ by $(\Lambda^s u)\hat{}\ (\xi) = (1 + |\xi|^2)^{\frac{s}{2}}\, \hat{u}(\xi)$. Show that $\Lambda^s: H^t(\mathbb{R}^n) \to H^{t-s}(\mathbb{R}^n)$ isomorphically. Using these operators, show that if $A \in PS(m,\rho,\delta)$ is properly supported then $A: H^s_{loc} \to H^{s-m}_{loc}$, if $\delta < \rho$.

In the proof of L^2 continuity given above, the assumption $\delta < \rho$ was necessary in order to ensure that the asymptotic series defining B was indeed an asymptotic series, i.e. that the order of the terms went to $-\infty$.

Recently, Calderon and Vaillancourt [88] have shown that $A \in PS(0,\rho,\rho)$ is continous on L^2. Beals and Fefferman [86] have made incisive use of this, in the case $\rho = ½$. For further generalizations, see [87]. Note that the proof of theorem 1 works for $A \in PS(0,0,0)$.

7. Families of pseudo differential operators.

We make $S^m_{\rho,\delta}(\Omega)$ into a Frechet space by means of the seminorms

$$|p|_{K,\alpha,\beta} = \sup_{x\in K} |p^{(\alpha)}_{(\beta)}(x,\xi)\,(1+|\xi|)^{-m+\rho|\alpha|-\delta|\beta|}|$$

where K is a compact subset of Ω, and α and β are multiindices.

We leave the following assertions as exercises for the reader. We shall make use of them later on, especially in Chapters V and VI.

1. The map $p \to p(x,D)$ is a continuous linear map from $S^m_{\rho,\delta}(\Omega)$ to $(H^s_{comp}(\Omega), H^{s-m}_{loc}(\Omega))$, if $\delta < \rho$.

2. Consider $p_\varepsilon(x,\xi) = e^{-\varepsilon|\xi|^2}$. If $\varepsilon > 0$, $p_\varepsilon \in S^{-\infty}_{1,0}(\mathbb{R}^n)$. Show that $\{p_\varepsilon : 0 \le \varepsilon \le 1\}$ is a <u>bounded</u> subset of $S^0_{1,0}(\mathbb{R}^n)$. In this case, what one must show is that

$$|D^\alpha_\xi\, p_\varepsilon| \le C_\alpha(1+|\xi|)^{-|\alpha|} \quad (0 \le \varepsilon \le 1)$$

where the constants C_α are independent of ε.

If X is a differentiable manifold which is paracompact, we give $PS(m,\rho,\delta;X)$ the topology induced from symbols obtained by local coordinate representations. We leave a precise formulation to the reader.

A Friedrichs' mollifier on X in a set of pseudo differential operators J_ε, $0 \leq \varepsilon \leq 1$, such that

(i) $J_\varepsilon \in PS(-\infty,1,0)$ for $\varepsilon > 0$, J_ε is properly supported.

(ii) $J_\varepsilon u \to u$ in L^2 for $u \in L^2_{comp}$, as $\varepsilon \longrightarrow 0$.

(iii) $\{J_\varepsilon : 0 \leq \varepsilon \leq 1\}$ is a bounded subset of $PS(0,1,0)$.

3. Using partitions of unity subordinate to coordinate coverings of X, and the result of exercise 2, show that X has a Friedrichs' mollifier.

4. Let $A \in PS(m,\rho,\delta;X)$, $1-\rho \leq \delta < \rho$. If J_ε is a Friedrichs' mollifier on X, show that $[A, J_\varepsilon] = AJ_\varepsilon - J_\varepsilon A$ has the following properties:

(i) $[A,J_\varepsilon] \in PS(-\infty,\rho,\delta)$ for $\varepsilon > 0$.

(ii) $\{[A,J_\varepsilon] : 0 \leq \varepsilon \leq 1\}$ is a bounded subset of $PS(m-(\rho-\delta),\rho,\delta)$.

5. For convenience, we take X to be compact in this exercise. Let $p(x,D) \in PS(1,1,0;X)$, $f \in L^2(X)$. A function $u \in L^2(X)$ is said to be a <u>weak</u> solution of the equation

$$p(x,D)\, u = f \qquad (*)$$

if this equation is valid, when $p(x,D)$ is applied to u in the distribution sense. On the other hand, it is said to be a <u>strong</u> solution of (*) if there exists a sequence $u_j \to u$ in $L^2(X)$ with $u_j \in C^\infty(X)$, such that $p(x,D)\, u_j = f_j \to f$ in $L^2(X)$. Clearly every strong solution of (*) is a weak solution.

Making use of the smooth functions $u_\varepsilon = J_\varepsilon u$, prove that, conversely, every weak solution of (*) is a strong solution.

This Friedrichs' mollifier technique was introduced by Friedrichs [23] in order to prove such "weak=strong" results. The technique has also proved a useful tool in passing from a priori estimate to regularity theorems as we will see in Chapter V. For weak=strong results on manifolds with boundary, see [71] and [67]. Such a result is also implicitly contained in [75].

8. Gårding's inequality.

This inequality, first proved by Gårding [26] for differential operators and by Calderon and Zygmund for singular integral operators goes far beyond being a tool to solve the Dirichlet problem for strongly elliptic operators. As a tool for proving energy inequalities for hyperbolic equations. (a task to which the tool was originally applied by Gårding [27]) and for proving other important *a priori* inequalities, it must rank as one of the fundamental results of the theory.

Theorem: Let $A \in PS(m,\rho,\delta)$ and assume $0 \leq \delta < \rho \leq 1$. Suppose that $\operatorname{Re} \sigma_A(x,\xi) \geq C|\xi|^m$ for ξ large, $C > 0$. Then for any real s, for any fixed compact set K, and all $u \in C_0^\infty(K)$, we have

(1) $\operatorname{Re}(Au,u) \geq C_0 \|u\|^2_{m/2} - C_1 \|u\|^2_s$, with C_0 and C_1

independent of u.

Proof: This is a consequence of the lemma proved in section 6, applied to the zero-order operator $p(x,D) = \Lambda^{-m/2} A \Lambda^{-m/2}$, since then $(Au,u) = (p(x,D) \Lambda^{m/2} u, \Lambda^{m/2}u)$.

There is also a sharp form of Gårding's inequality which says that under the weaker hypothesis that $\operatorname{Re} \sigma_A(x,\xi) \geq 0$, if $A \in PS(m,1,0,)$ it follows that, for K compact,

$$\operatorname{Re}(Au,u) \geq -C_1 \|u\|^2_{(m-1)/2} \qquad u \in C_0^\infty(K).$$

We shall return to this in Chapters VI and VII.

Chapter III. Elliptic and Hypoelliptic Operators

1. Elliptic operators.

Definition: $p(x,D) \in PS(m,\rho,\delta;X)$ is elliptic of order m if for each compact $K \subset X$, there are constants C_K and R such that

$$|p(x,\xi)| \geq C_K(1+|\xi|)^m \quad \text{if} \quad |\xi| \geq R, \quad x \in K.$$

Then the Laplace operator $\Delta = \frac{\partial^2}{\partial x_1^2} + \cdots + \frac{\partial^2}{\partial x_n^2}$ is elliptic of order two, since $\sigma_\Delta(x,\xi) = -\xi_1^2 - \cdots - \xi_n^2 = -|\xi|^2$, but the wave operator $\Box = \Delta_{n-1} - \frac{\partial^2}{\partial x_n^2}$, with symbol $\sigma_\Box(x,\xi) = -|\xi'|^2 + \xi_n^2$, is not elliptic.

Definition: A parametrix P for the operator $Q \in PS(m,\rho,\delta)$ is a pseudo-differential operator which is a right and left inverse for Q modulo smoothing operators:

$$PQ - I = K_1 \in PS(-\infty)$$

$$QP - I = K_2 \in PS(-\infty) .$$

Exercise 1. If $q(x,D)$ is elliptic of order m, $q(x,D) \in PS(m,\rho,\delta)$, show that $q(x,\xi)^{-1} \in S^{-m}_{\rho,\delta}$ for large ξ.

THEOREM: If $q(x,D) \in PS(m,\rho,\delta)$ is elliptic, $\delta < \rho$, then there is a properly supported $P \in PS(-m,\rho,\delta)$ which is a parmetrix of $q(x,D)$.

Proof. We will write the symbol p of $P = p(x,D)$ as an asymptotic su

$p \sim \Sigma p_j$ and get successive approximations to p. First, set

$$p_0(x,\xi) = \zeta(x,\xi)\, q(x,\xi)^{-1} \in S^{-m}_{\rho,\delta} ,$$

where ζ vanishes in a nbd of the zeros of q, and is identically 1 for large ξ. The formula for a product yields

$$\sigma_{P_0 Q} \sim \Sigma \frac{1}{\alpha!}\, p_0^{(\alpha)}(x,\xi)\, q_{(\alpha)}(x,\xi) = 1 + S_{-1}(x,\xi)$$

where $S_{-1} \in S^{m-(\rho-\delta)}_{\rho,\delta}$. Similarly, we have

$$\sigma_{(P_0 + P_1)Q} \sim 1 + S_{-1}(x,\xi) + \sum_{\alpha\ 0} \frac{1}{\alpha!}\, p_1^{(\alpha)}(x,\xi)\, q_{(\alpha)}(x,\xi) + p_1(x,\xi)\, q(x,\xi) .$$

Then we take $p_1(x,\xi) = -S_{-1}(x,\xi)\, q(x,\xi)^{-1}$, for large ξ, yielding

$$\sigma_{(P_0+P_1)Q} = 1 + S_{-2} ; \quad S_{-2} \in S^{m-2(\rho-\delta)}_{\rho,\delta} .$$

In general, if at stage j we have

$$\sigma_{(P_0 + \cdots + P_{j-1})Q} = 1 + S_{-j}(x,\zeta) , \quad S_{-j} \in S^{m-j(\rho-\delta)}_{\rho,\delta} ,$$

set $p_j(x,\xi) = -S_{-j}(x,\zeta)\, q(x,\xi)^{-1}$ for large j, and continue. It is easy to verify that $p \sim \Sigma p_j$ is a right parametrix of Q, i.e. $PQ - I = K_1 \in PS(-\infty)$.

Similarly, one can construct a left parametrix $P^1 \in PS(-m,\rho,\delta)$ such that $QP^1 - I = K_2 \in PS(-\infty)$. However, note that

$$PQP^1 = (1 + K_1)P^1 = P^1 + K_1 P^1 \quad \text{and}$$

$$PQP^1 = P(1 + K_2) = P + PK_2 .$$

Hence $P - P^1 = K_1 P^1 - PK_2 \in PS(-\infty)$, so P itself is a two-sided parametrix of Q.

As a consequence, we have the following regularity theorem.

Corollary. Let $Q \in PS(m, \rho, \delta)$ be elliptic. If $u \in \mathcal{E}'(\Omega)$ and if $Qu|_\omega \in C^\infty(\omega)$, ω an open subset of Ω, then $u|_\omega$ is smooth.

Proof: If P is a properly supported parametrix of Q, we can assume further that the restriction of the distribution kernel of P to $\omega \times \omega$ is properly supported. Then if $f = Qu|_\omega \in C^\infty(\omega)$, we have $Pf|_\omega = PQu|_\omega = u + Ku|_\omega$. Since $Pf \in C^\infty(\omega)$ and K is smoothing $u|_\omega \in C^\infty(\omega)$, as asserted.

The crucial propery of the parametrix used in the proof of this corollary is that its distribution kernel is smooth off the diagonal.

§ 2. Hypoelliptic operators with constant strength.

In the last section, we showed that elliptic operators satisfy a certain smoothness property. There is a more general class of operators which satisfy such a property, of which the **heat operator** $\frac{\partial}{\partial t} - \Delta$ is another important example, which we now discuss.

Definition. A properly supported pseudo-differential operator $P \in PS(m, \rho, \delta; \Omega)$ is hypoelliptic if for any open $\omega \subset \Omega$, $Pu|_\omega \in C^\infty(\omega)$ implies $u|_\omega \in C^\infty(\omega)$, for any $u \in D'(\Omega)$.

THEOREM 1. Let $P(\xi)$ be a polynomial. The following are equivalent.

(1) $P(D)$ is hypoelliptic.

(2) $$\left| \frac{P^{(\alpha)}(\xi)}{P(\xi)} \right| \leq C(1 + |\xi|)^{-\rho|\alpha|} \quad \text{for } |\xi| \text{ large.}$$

(3) If $V = \{\zeta \in \mathbb{C}^n : P(\zeta) = 0\}$ then $|\text{Re}\,\zeta| \to \infty$, $\zeta \in V \Rightarrow |\text{Im}\,\zeta| \to \infty$

(4) There exists $\rho > 0$, $C > 0$, such that for $\zeta \in V$, $|\zeta|$ large, $|\text{Im}\,\zeta| \geq C|\zeta|^\rho$.

Proof. First we show that the inequality of part (2) implies that $P(D)$ is hypoelliptic. In fact, we have done most of the work needed to prove this in Chapter 1. The reader should verify that, if (2) is satisfied, then $P(\xi)^{-1} \in S^0_{\rho,0}$ for large ξ. The proof that (2) $\Rightarrow$ (1) is then easily completed by arguments we've seen in the last section.

Next we show that (1) implies (3). For this, let

$$\mathcal{N} = \{u \in C(\Omega) : P(D)u = 0\} \subset C(\Omega)$$

$$= \{u \in C^1(\Omega) : P(D)u = 0\} \subset C^1(\Omega), \text{ if } P(D) \text{ hypoellipti}$$

Since $\mathcal{N}$ is a Frechet space under topologies induced from either $C(\Omega)$ or $C^1(\Omega)$, and since one is stronger than the other, the open mapping theorem implies these two topologies coincide.

Thus, if $\zeta_\nu \in V$, $\{\operatorname{Im} \zeta_\nu\}$ bounded, then $\{e^{i\langle \zeta_\nu, x\rangle}\}$ is a subset of $\mathcal{N}$, bounded in $C(\Omega)$. The conclusion is:

$$\{e^{i\langle \zeta_\nu, x\rangle}\} \text{ is bounded in } C^1(\Omega).$$

Since $\frac{\partial}{\partial x_j} e^{i\langle \zeta_\nu, x\rangle} = i\zeta_\nu^j e^{i\langle \zeta_\nu, x\rangle}$, where $\zeta_\nu = (\zeta_\nu^1, \cdots, \zeta_\nu^n)$, it follows that $\{\zeta_\nu\}$ is bounded, and the implication (1) $\Rightarrow$ (3) follows.

It remains to show that (2), (3), and (4) are all equivalent. That (3) $\Leftrightarrow$ (4) can be deduced from the Seidenberg-Tarski theorem. We refer the reader to Hörmander [31] or Treves [80] for this argument. It remains to show that (2) $\Leftrightarrow$ (4), and this is an easy consequence of the following lemma.

<u>Lemma 1.</u> Let $P(\xi)$ be a polynomial and let $d(\xi) = \operatorname{dist}(\xi, V)$, the distance from $\xi \in \mathbb{R}^n$ to $V = \{\zeta \in \mathbb{C}^n : P(\zeta) = 0\}$. Then

$$C \leq d(\xi) \sum_{\alpha \neq 0} \left| \frac{P^{(\alpha)}(\xi)}{P(\xi)} \right|^{\frac{1}{|\alpha|}} \leq C'.$$

<u>Proof.</u> $P(\xi + \eta) - P(\xi) = \sum_{\alpha \neq 0} \frac{\eta^\alpha}{\alpha!} P^{(\alpha)}(\xi)$. Hence

$$(*) \qquad \frac{P(\xi+\eta)}{P(\xi)} - 1 = \sum_{\alpha \neq 0} \frac{\eta^\alpha}{\alpha!} \frac{P^{(\alpha)}(\xi)}{P(\xi)} .$$

Pick C such that $\sum_{\alpha \neq 0} \frac{C^{|\alpha|}}{\alpha!} < 1$. Suppose $|\eta| < C \min_{\alpha \neq 0} \left| \frac{P^{(\alpha)}(\xi)}{P(\xi)} \right|^{-\frac{1}{|\alpha|}}$,

Then (*) implies that $P(\xi + \eta) \neq 0$. Hence

$$d(\xi) \geq C \min_{\alpha \neq 0} \left| \frac{P^{(\alpha)}(\xi)}{P(\xi)} \right|^{-\frac{1}{|\alpha|}} ,$$ which is half the lemma.

For the second half, let $|\eta| \leq d(\xi)$. Then if $g(t) = P(\xi + t\eta)$, we have

$$\left| \frac{P(\xi+\eta)}{P(\xi)} \right| = \left| \frac{g(1)}{g(0)} \right| = \prod_{j=1}^{m} \left| \frac{1 - t_i}{t_j} \right| \leq 2^m$$ where t_j are the

roots of $g(t)$; $|t_j| \gtrsim 1$. Thus $|P(\xi+\eta)| \leq |2^m P(\xi)|$ if $|\eta| \leq d(\xi)$.

If we apply this to Cauchy's formula

$$P^{(\alpha)}(\xi) = C \int_{\Gamma_n} \frac{P(\xi + \eta)}{\zeta^\alpha \zeta_1 \cdots \zeta_n} d\zeta , \quad \text{where} \quad \Gamma_n = 2^{-\frac{1}{2n}} d(\xi) \mathbb{T}^n ,$$

we get $|P^{(\alpha)}(\xi)| \leq C \frac{2^m |P(\xi)|}{d(\xi)^{|\alpha|}}$, which completes the proof.

Thus we have completely characterized those hypoelliptic differential operators with constant coefficients. For operators with variable coefficients, results are more subtle and less complete. There are several directions to go in, looking for large families of hypoelliptic operators. Here, we examine one of them.

<u>Definition:</u> $p(x, D) = \sum_{|\alpha| \leq k} a_\alpha(x) D^\alpha$ is a formally hypoelliptic operator of constant strength if there exists a $P(D)$ such that

(i) $P(D)$ is hypoelliptic, and

(ii) for each fixed x, $\frac{P(x, \xi)}{P(\xi)}$ and $\frac{P(\xi)}{P(x, \xi)}$ are bounded, for large ξ

Proposition 1: If $P(D)$ is hypoelliptic and if $Q(\xi)$ is a polynomial such that $\frac{P(\xi)}{Q(\xi)}$ and $\frac{Q(\xi)}{P(\xi)}$ are bounded, for large ξ, then $Q(D)$ is also hypoelliptic.

<u>Proof:</u> Let $d_1(\xi)$ be the distance from ξ to $V = \{\zeta \in \mathbb{C}^n : P(\zeta) = 0\}$ and let $d_2(\xi)$ be the corresponding distance function to the variety of zero of $Q(\xi)$. The above lemma yields

$$(1) \qquad d_1(\xi)^{|\alpha|} \, |P^{(\alpha)}(\xi)| \leq C^{|\alpha|} \, |P(\xi)| \, .$$

To complete the proof, we shall need the following fact, which we shall prove shortly: if $\frac{Q(\xi)}{P(\xi)}$ is bounded for large ξ, then with $\tilde{R}(\xi, t) = \left(\sum_\alpha |R^{(\alpha)}(\xi)|^2 \, t^{2|\alpha|}\right)^{\frac{1}{2}}$ for any polynomial $R(\xi)$,

$$(2) \qquad \tilde{Q}(\xi, t) \leq C \, \tilde{P}(\xi, t) \quad \text{for} \quad \xi \in \mathbb{R}^n, \ t \geq 1 \, .$$

Now, when ξ is so large that $d_1(\xi) \geq 1$, we apply inequality (2) to

$t = d_1(\xi)$, and use inequality (1) to get

$$\sum_\alpha |Q^{(\alpha)}(\xi)|^2 \, d_1(\xi)^{2|\alpha|} \leq C|P(\xi)|^2 .$$

Since $\frac{P(\xi)}{Q(\xi)}$ is bounded for large ξ, we have

$$\Sigma \, |Q^{(\alpha)}(\xi)|^2 d_1(\xi)^{2|\alpha|} \leq C|Q(\xi)|^2 .$$

Since $d_1(\xi) \geq C'|\xi|^\rho$ for large ξ, with $\rho > 0$, we immediately see that Q satisfies condition (2) of Theorem 1; hence, $Q(D)$ is hypoelliptic.

It remains to prove the inequality (2). We shall obtain this as the second of the following two lemmas.

Lemma 2: There are constants C_1, C_2 such that, for all polynomials R of degree $\leq K$, we have, with $\zeta \in \mathbb{R}^n$, $\xi \in \mathbb{R}^n$, $t \geq 0$,

$$C_1 \tilde{R}(\xi, t) \leq \sup_{|\zeta| \leq t} |R(\xi + \zeta)| \leq C_2 \tilde{R}(\xi, t) .$$

Proof: The right-hand inequaltiy follows from the Taylor expansion of $R(\xi + \zeta)$ about ξ. Since, if we set $R_t(\xi) = R(t\xi)$, we get $\tilde{R}(\xi, t) = \tilde{R}_t(\xi/t, 1)$, it suffices to prove the other inequaltiy with $t = 1$.

Pick N distinct points $\zeta_1, \cdots, \zeta_N$ in the unit ball in $\mathbb{R}^n$, where N is the dimension of the space of polynomials in $\mathbb{C}[X_1, \cdots, X_n]$ of degree $\leq K$. Then the equations for the coefficient of polynomials P_j, given by

$$P_j(\zeta_k) = \delta_{jk}$$

can be solved uniquely for polynomials of degree $\leq K$, and we obtain

$$R(\eta) = \sum_{j=1}^{N} R(\zeta_j) P_j(\eta)$$

known as the Lagrange interpolation formula.

$$\therefore \quad R(\xi + \eta) = \sum_{j=1}^{N} R(\xi + \zeta_j) P_j(\eta)$$

$$\therefore \quad R^{(\alpha)}(\xi) = \sum_{j=1}^{N} P_j^{(\alpha)}(0) R(\xi + \zeta_j) .$$

The left-hand inequality, with $t = 1$, now follows easily.

Lemma 3: Inequality (2) above is valid: $\tilde{Q}(\xi, t) \leq C \tilde{P}(\xi, t)$ for $t \geq 1$, all ξ

Proof: Using Lemma 2, and the boundedness of $\frac{Q(\xi)}{P(\xi)}$ for large ξ,

$$\begin{aligned} Q(\xi, t) &\leq C_1 \sup_{|\zeta| \leq t} |Q(\xi + \zeta)| \\ &\leq C_2 (1 + \sup_{|\zeta| \leq t} |P(\xi + \zeta)|) \\ &\leq C_2 \sup_{|\zeta| \leq t} \tilde{P}(\xi + \zeta, 1) \\ &\leq C_3 \sup_{|\zeta| \leq t+1} P(\xi + \zeta) \leq C_4 \tilde{P}(\xi, t+1) \\ &\leq C_4 (1 + \tfrac{1}{t})^m \tilde{P}(\xi, t) . \end{aligned}$$

Thus the proof of proposition 1 is complete. In particular, it follows that if $p(x, D)$ is formally hypoelliptic of constant strength, then any constant coefficient operator $P_0(D) = p(x_0, D)$ obtained by freezing the coefficients of $p(x, D)$ at some point, is hypoelliptic.

The following corollary of proposition 1 will be of interest.

Lemma 4: If $P(D)$ is hypoelliptic and $\frac{Q(\xi)}{P(\xi)}$ is bounded for large ξ, then

$$\left| \frac{Q^{(\alpha)}(\xi)}{P(\xi)} \right| \leq C|\xi|^{-\rho|\alpha|} \qquad \text{for large } \xi .$$

Proof: Pick ε so small that $C_1 P(\xi) \leq P(\xi) + \varepsilon Q(\xi) \leq C_2 P(\xi)$ for large ξ. By proposition 1, $P_1(D) = P(D) + \varepsilon Q(D)$ is hypoelliptic. Hence we have

$$\frac{Q^{(\alpha)}(\xi)}{P(\xi)} = \frac{1}{\varepsilon} \frac{P_1^{(\alpha)}(\xi) - P^{(\alpha)}(\xi)}{P(\xi)}$$

$$= \frac{1}{\varepsilon} \frac{P_1^{(\alpha)}(\xi)}{P_1(\xi)} \frac{P_1(\xi)}{P(\xi)} - \frac{1}{\varepsilon} \frac{P^{(\alpha)}(\xi)}{P(\xi)}$$

$$\leq C|\xi|^{-\rho|\alpha|} \qquad \text{for } \xi \text{ large.}$$

Suppose now that $p(x, D)$ is a formally hypoelliptic operator with constant strength. Freeze the coefficients at one point x_0 to obtain a hypoelliptic constant coefficient operator $P(D) = p(x_0, D)$. Let E be a properly supported pseudo-differential operator, $E \in PS(0, \rho, 0)$, such that $\sigma_E(x, \xi) = P(\xi)^{-1}$ for large ξ. As we have seen in the proof of Theorem 1,

E is a parametrix of $P(D)$.

Proposition 2: $p(x, D)E$ and $E\,p(x, D)$ belong to $PS(0, \rho, 0)$ and are elliptic.

Proof: That $p(x, \xi)\,P(\xi)^{-1} \in S^{0}_{\rho, 0}$ for large ξ is a routine estimation, using Lemma 4. From this we have $p(x, D)E \in PS(0, \rho, 0)$. That $E\,p(x, D) \in PS(0, \rho, 0)$ follows from looking at the terms in the asymptotic expansion of its symbol, and we leave the details to the reader. That the operators are elliptic is obvious.

Now for the second main result of this section.

THEOREM 2: If $p(x, D)$ is a formally hypoelliptic operator of constant strength, then $p(x, D)$ is hypoelliptic.

Proof: If $A \in PS(0, \rho, 0)$ is a parametrix for the elliptic operator $E\,p(x, D)$, then $(AE)\,p(x, D) - I \in PS(-\infty, \rho, 0)$, so AE is a left parametrix of $p(x, D)$, and the argument of the previous section shows that $p(x, D)$ has the required regularity property.

1. Let $A(x, D) = \sum_{|\alpha| \leq 2} a_\alpha(x) D^\alpha$ be a second order operator, with

$D_j = \frac{1}{i} \frac{\partial}{\partial x_j}$, which is strongly elliptic in the sense that

$$\text{Re} \sum_{|\alpha|=2} a_\alpha(x)\, \xi^\alpha \geq C|\xi|^2 .$$

We assume the coefficients are smooth; they could also depend on t. Prove that $\frac{\partial}{\partial t} - A(x, D)$ is formally hypoelliptic with constant strength

2. Consider the hypoelliptic operator $P(D) = \frac{\partial}{\partial t} - \frac{\partial^2}{\partial x^2}$ on $\mathbb{R}^2$. Prove that there exists a non-linear change of coordinates of $\mathbb{R}^2$ with respect to which P becomes an operator which is not formally hypoelliptic. Is this new operator hypoelliptic ?

3. Let X and Y be two smooth vector fields in $\mathbb{R}^2$ which are linearly independent at each point. Prove that the second order operator $X^2 - Y$ is hypoelliptic.

4. Let $p(x, D)$ be a formally hypoelliptic operator with constant strength. Prove that its adjoint $p(x, D)^*$ also enjoys this property.

In general, the adjoint of a hypoelliptic operator has the local solvability property. See Treves [77] ; a similar argument, in another context, is given in Chapter VI of these notes.

§3. References to further work.

There is a more general notion of operators of constant strength than that presented here; see [31] . It is known ([76]) that every hypoelliptic operator of constant strength is formally hypoelliptic. However, there are other hypoelliptic operators which are not of this sort. Treves and Hörmander have examined a class of operators that one may call formally hypoelliptic with slowly varying strength. In fact, the classes $PS(m, \rho, \delta)$ with $\delta > 0$, which we have not used, were invented to handle these operators; see [34] .

There is a nearly complete characterization of second order hypoelliptic operators with real coefficients. See [36] and, for a simpler treatment using pseudo-differential operators, see [43] . This topic is also treated in [63] .

There is also a theory of hypoelliptic operators of principal type, for which we refer the reader to [79] .

Chapter IV. The Initial Value Problem. Hyperbolic Operators

In this chapter our main aim is to study the Cauchy problem for hyperbolic operators. That is, if $L(x, t, D)$ is a hyperbolic operator of order m, we wish to solve the initial value problem

$$\begin{aligned} Lu &= f \\ u(x, 0) &= g_1 \\ \frac{\partial}{\partial t} u(x, 0) &= g_2 \\ &\;\vdots \\ \frac{\partial^{m-1}}{\partial t^{m-1}} u(x, 0) &= g_m \,. \end{aligned}$$

We assume $L = \dfrac{\partial^m}{\partial t^m} - \sum_{j=0}^{m-1} A_{m-j}(x, t, D_x) \dfrac{\partial^j}{\partial t^j}$ where A_{m-j} is a differential operator in the x variables of order $m-j$, depending smoothly on the parameter t. The symbol of L is

$$L(x, t, \xi, \tau) = (i\tau)^m - \sum_{j=0}^{m-1} A_{m-j}(x, t, \xi)(i\tau)^j \,.$$

There are several notions of what it means for such an operator to be hyperbolic. We shall introduce a couple of them in later sections, as we attempt to solve the above initial value problem.

§1. Reduction to a first order system.

Suppose the coefficients of L are smooth and defined on all of $\mathbb{R}^n$. In that case, with $(\Lambda u)^\wedge(\xi) = (1+|\xi|^2)^{1/2}\,\hat{u}(\xi)$, write

$$u_1 = \Lambda^{m-1} u$$

$$u_2 = \frac{\partial}{\partial t}\Lambda^{m-2} u$$

$$\vdots$$

$$u_j = \left(\frac{\partial}{\partial t}\right)^{j-1}\Lambda^{m-j} u$$

$$\vdots$$

$$u_m = \frac{\partial^{m-1}}{\partial t^{m-1}} u$$

Then the equation $Lu = f$ turns into the first order system

$$\frac{\partial}{\partial t}\begin{pmatrix} u_1 \\ \cdot \\ \cdot \\ \cdot \\ u_m \end{pmatrix} = \begin{pmatrix} 0 & \Lambda & 0 & \cdots & 0 \\ 0 & 0 & \Lambda & \ddots & 0 \\ \cdot & \cdot & \cdot & \ddots & \cdot \\ \cdot & \cdot & \cdot & & \Lambda \\ b_1 & b_2 & b_3 & \cdots & b_m \end{pmatrix}\begin{pmatrix} u_1 \\ \cdot \\ \cdot \\ \cdot \\ u_m \end{pmatrix} + \begin{pmatrix} 0 \\ 0 \\ \cdot \\ 0 \\ f \end{pmatrix} = K\begin{pmatrix} u_1 \\ \cdot \\ \cdot \\ \cdot \\ u_m \end{pmatrix} + \begin{pmatrix} 0 \\ 0 \\ \cdot \\ 0 \\ f \end{pmatrix}$$

where $b_j = A_{m-j+1}(x, t, D_x)\,\Lambda^{j-m}$. Now $K \in PS(1,1,0)$ and it has a principal symbol, which we call $\sigma_{K_0}(x, t, \xi)$ which is a matrix valued function, homogeneous of degree 1 in ξ. Furthermore, as the reader can verify, the eigenvalues of $\sigma_{K_0}(x, t, \xi)$ differ from the roots $\tau_1(\xi), \cdots, \tau_m(\xi)$ of $L_m(x, t, \xi, \tau) = 0$ by a factor of i, where L_m is the principal part of L, consisting of all the terms of L of order precisely m.

It is clear that specifying the Cauchy data $\frac{\partial^j}{\partial t^j} u(x, 0)$, $j=0, \cdots, m-1$ of u is equivalent to specifying $u_1, \cdots, u_m$ at $t = 0$.

Now, to see when we can solve the initial value problem

$$\frac{\partial}{\partial t} u = Ku + f$$

$$u(x, 0) = g$$

with $K \in PS(1, 1, 0)$ a matrix-valued operator, let us take a look at the constant coefficient case, $\sigma_K(x, t, \xi) = k(\xi)$. With $f = 0$, $\hat{u}(\xi, t) = (2\pi)^{-n} \int u(x, t) e^{i\langle x, \xi \rangle} dx$, we have

$$\hat{u}(\xi, t) = e^{t k(\xi)} \hat{g}(\xi).$$

If we require that the initial datum $g \in L^2(\mathbb{R}^n)$ leads to $u(\cdot, t) \in L^2(\mathbb{R}^n)$ for each $t \in \mathbb{R}$, we need $e^{t k(\xi)}$ bounded in ξ, for each real t. If $k(\xi)$ is homogeneous of degree 1 in ξ, this is equivalent to the assertion that, for each $\xi \neq 0$, $k(\xi)$ is diagonalizable, and all its eigenvalues are pure imaginary.

If we recall how the roots of $L_m(x, t, \xi, \tau) = 0$ are connected to the eigenvalues of $\sigma_{K_0}(x, t, \xi)$, the following definition becomes reasonable.

<u>Definition:</u> L is hyperbolic if the roots $\tau_1(x, t, \xi), \cdots, \tau_m(x, t, \tau)$ of $L_m(x, t, \xi, \tau) = 0$ are all real.

Hyperbolicity in the broad sense will not allow us to solve the Cauchy problem, but in the next two sections we introduce further conditions which will.

In the next two sections, we shall suppose, for technical convenience, that the coefficients of all operators are periodic with respect to the x variables, so that our unknown solution u is to be defined, say, in $[0, \tau] \times \mathbb{T}^n$, where $\mathbb{T}^n$ is the n-torus.

The main point of sections 2 and 3 is just to get local existence theorems We get good global existence theorems by the finite propagation speed/ finite domain of dependence argument of section 4.

§2. Symmetric hyperbolic systems.

Definition: The operator $\frac{\partial}{\partial t} - K$ is symmetric hyperbolic if $K + K^* \in PS(0,1,0)$, where $K = K(t)$ is a smooth one-parameter family of operators in $PS(1,1,0)$.

We shall prove that a solution of the initial value problem exists and is unique by proving an a priori inequality for such solutions. Existence will then follow, by a little functional analysis. The following inequality from the the theory of ordinary differential equations will be a useful tool.

Lemma (Gronwall's inequality): If $y \in C^1$ and $y'(t) + f(t)y \leq g(t)$, then

$$(1) \qquad y(t) \leq e^{-\int_0^t f(\tau)\, d\tau} \left[y_0 + \int_0^t g(\tau)\, e^{\int_0^\tau f(\lambda)\, d\lambda}\, d\tau \right].$$

Proof: The hypothesis is equivalent to the inequality

$$\frac{d}{dt}\left(y\, e^{\int_0^t f} \right) \leq g(t)\, e^{\int_0^t f}$$

and integrating this yields (1).

Suppose now that $u \in C_0^\infty(\mathbf{T}^{n+1})$ and $\frac{\partial}{\partial t} u - Ku = f$. If we differentiate $\| u(t) \|^2 = \int | u(x,t)|^2\, dx$, we respect to t, we get

$$\begin{aligned} \frac{d}{dt}(u,u) &= (u', u) + (u, u') \\ &= (Ku + f, u) + (u, Ku + f) \\ &= ((K + K^*)u, u) + 2\,\mathrm{Re}\,(f,u) \\ &\leq C\|u\|^2 + C\|f\|^2, \text{ since } K + K^* \in PS(0,1,0) \end{aligned}$$

Applying Gronwall's inequality to this yields

(2) $$\| u(t) \|^2 \leq C' \| u(0) \|^2 + C' \int_0^T \| f(\tau) \|^2 \, d\tau , \quad 0 \leq t \leq T ,$$

where $C' = C'(T)$ is independent of u, f, and t, for $0 \leq t \leq T$.

Exercise 1: Differentiating $\| \Lambda^S u(t) \|^2$, prove that, for all real S,

(3) $$\| u(t) \|_S^2 \leq C \| u(0) \|_S^2 + C \int_0^t \| f(\tau) \|_S^2 \, d\tau , \quad 0 \leq t \leq T ,$$

where $\| \ \|_S$ is the norm in $H^S(\mathbb{T}^n)$, given $u \in C_0^\infty(\mathbb{R} \times \mathbb{T}^n)$, $u' - Ku = f$

Hint: $[\Lambda^S, K] \in PS(S, 1, 0)$.

More generally, (2) and (3) are valid for $u \in C^1([0,T], H^{S+1}(\mathbb{T}^n))$, $f \in C([0,T], H^S(\mathbb{T}^n))$, $\varphi_0 \in H^{S+1}(\mathbb{T}^n)$.

We shall obtain the solution of the initial value problem

(*) $$\frac{\partial}{\partial t} u = Ku + f$$
$$u(0) = \varphi_0$$

as a limit of solutions to the problem

(**) $$\frac{\partial}{\partial t} u = K J_\varepsilon u + f$$
$$u(0) = \varphi_0$$

where J_ε is a Friedrichs' mollifer in the x-directions, say $(J_\varepsilon g)^\wedge(\xi) = e^{-t|\xi|^2} \hat{g}(\xi)$. The point of this is that, for each $\varepsilon > 0$, $K_\varepsilon = KJ_\varepsilon$ in a continuous operator in $H^S(\mathbb{T}^n)$. Then (**) can be regarded as a Banach-space valued ordinary differential equation, to which the Picard iteration method applies (see [15]). Then, given $\varphi_0 \in H^S(\mathbb{T}^n)$, $f \in C([0, T], H^S(\mathbb{T}^n))$, we can solve (**), producing a solution $u_\varepsilon \in C^1([0,T], H^S(\mathbb{T}^n))$.

Furthermore, $K_\varepsilon : 0 \leq \varepsilon \leq 1$ is a bounded subset of $PS(1, 1, 0)$ and $\{K_\varepsilon + K_\varepsilon^* : 0 \leq \varepsilon \leq 1\}$ is a bounded subset of $PS(0, 1, 0)$.

The crucial implication of this is that the energy inequality

$$(4) \qquad \| u_\varepsilon(t) \|_S^2 \leq C \| \varphi_0 \|_S^2 + C \int_0^t \| f(\tau) \|_S^2 \, d\tau , \quad 0 \leq t \leq T$$

holds with constant C independent of ε, $0 < \varepsilon \leq 1$. Now, of course, we want to let $\varepsilon \to 0$.

First note that, by inequality (4), $\{u_\varepsilon : 0 < \varepsilon \leq .1\}$ is a bounded subset $C([0, \tau], H^S(\mathbb{T}^n))$, given $\varphi_0 \in H^S(\mathbb{T}^n)$. Since $u_\varepsilon' = K_\varepsilon u_\varepsilon + f$, it follows that $\{u_\varepsilon' : 0 < \xi \leq 1\}$ is a bounded subset of $C([0, \tau], H^{S-1}(\mathbb{T}^n))$; hence $\{u_\varepsilon\}$ is a bounded subset of $C^1([0, T], H^{S-1}(\mathbb{T}^n))$. Furthermore, for each $t_0 \in [0,T]$, $\{u_\varepsilon(t_0) : 0 < \varepsilon \leq 1\}$, being a bounded subset of $H^S(\mathbb{T}^n)$, is a relatively compact subset of $H^{S-1}(\mathbb{T}^n)$. Hence, by Ascoli's theorem (see [15] or [17]) there is a sequence $\varepsilon_n \to 0$ such that u_{ε_n} converges, in $C([0, \tau], H^{S-1}(\mathbb{T}^n))$, to a limit we call u. Clearly u satisfies (*), in the distribution sense.

Before we complete the last step of the argument, let us state precisely what we're proving.

THEOREM: Let $\frac{\partial}{\partial t} - K$ be a first order symmetric hyperbolic system. Then, given $\varphi_0 \in H^S(\mathbb{T}^n)$ and $f \in C([0,\tau], H^S(\mathbb{T}^n))$, the Cauchy problem (*) has a unique solution $u \in C([0,\tau], H^S(\mathbb{T}^n))$.

Proof: We have obtained a solution u to (*), with $u \in C([0,\tau], H^{S-1}(\mathbb{T}^n))$. Now let $\varphi_j \in H^{S+3}(\mathbb{T}^n)$, $f_j \in C([0,\tau], H^{S+3}(\mathbb{T}^n))$, be such that $\varphi_j \to \varphi_0$ in $H^S(\mathbb{T}^n)$ and $f_j \to f$ in $C([0,\tau], H^S(\mathbb{T}^n))$.

The arguement above shows that the associated solutions u_j of (*), with f, φ_0 replaced by f_j, φ_j, belong to $C([0,\tau], H^{S+2}(\mathbb{T}^n))$. Since $u_j' = Ku_j + f_j$, it follows that $u_j \in C^1([0,\tau], H^{S+1}(\mathbb{T}^n))$. Hence we can apply the energy inequality (3) and conclude that $\{u_j\}$ is a Cauchy sequence in $C([0,\tau], H^S(\mathbb{T}^n))$. Hence u belongs to this space, as asserted.

Since any $u \in C([0,\tau], H^S(\mathbb{T}^n))$ solving this Cauchy problem must belong to $C^1([0,\tau], H^{S-1}(\mathbb{T}^n))$, the energy inequality (3) applies, and we see that the solution is unique.

Exercise: Show that you only need $f \in L^2([0,\tau], H^s(\mathbb{T}^n))$.

§3. Strictly hyperbolic equations.

Definition: If $K = K_1 + K_0$ where $K_0 \in PS(0,1,0)$ and $\sigma_{K_1}(x,\xi)$ is homogeneous of degree 1 in ξ for $|\xi| \geq 1$, whose eigenvalues, for each fixed (x,ξ), $|\xi| \geq 1$, are pure imaginary and distinct, the operator $\frac{\partial}{\partial t} - K$ is called strictly hyperbolic.

In analogy, we have the following.

Definition: If L is a differential operator of order m, L is said to be strictly hyperbolic with respect to the initial surfaces t = const. if, for each fixed (x,t,ξ), the roots $\tau_1, \cdots, \tau_m$ of $L_m(x,t,\xi,\tau) = 0$ are real and distinct, provided $\xi \neq 0$.

The way we treat the Cauchy problem

$$\frac{\partial}{\partial t} u = Ku + f$$

$$u(0) = \varphi_0$$

when $\frac{\partial}{\partial t} - K$ is strictly hyperbolic is to construct what is called a symmetrizer for K, which will allow us to prove energy inequalities (2) and (3) of the last section.

Definition: A symmetrizer for $\frac{\partial}{\partial t} - K$ is a smooth one-parameter family of pseudo-differential operators $R = R(t) \in PS(0,1,0)$ such that

(i) $\sigma_R(x,\xi)$ is a positive definite matrix function, for $|\xi| \geq 1$.

(ii) $\sigma_{RK} = -\sigma_{RK}^{*} \mod S^0_{1,0}(\mathbb{T}^n)$.

Proposition: Any strictly hyperbolic first order system $\frac{\partial}{\partial t} - K$ has a symmetrizer.

Proof: If we let the eigenvalues of $\sigma_{K_0}(x,t,\xi)$, for $|\xi| \geq 1$, be $i\lambda_\nu(x,t,\xi)$, where $\lambda_1(x,t,\xi) < \lambda_2(x,t,\xi) < \cdots < \lambda_k(x,t,\xi)$, K being a k-by-k system, then λ_ν are well defined C^∞ functions of (x,t,ξ), homogeneous of degree 1 in ξ. Hence $\lambda_\nu \in S^1_{1,0}(\mathbb{T}^n)$ for large ξ.

Similarly, if $P_\nu(x,t,\xi)$ are the projections onto the associated eigenvalues of $i\lambda_\nu(x,t,\xi)$, $P_\nu = \frac{1}{2\pi i}\int_{\gamma_\nu} (\varphi - \sigma_{K_0}(x,t,\xi))^{-1}\, d\zeta$, then $P_\nu \in S^0_{1,0}(\mathbb{T}^n)$ for large ξ. Now let $R(x,t,D) = \sum_{j=1}^{k} P_j(x,t,D)^* P_j(x,t,D)$.

It is clear that R is a symmetrizer, and $R \in PS(0,1,0)$. Note that

$\sigma_{RK}(x,t,\xi) = i\sum_{j=1}^{k} \lambda_j(x,t,\xi)\; P_j(x,t,\xi)^* P_j(x,t,\xi) \mod S^0_{1,0}$.

With such a symmetrizer, we can prove an energy inequality. In fact, let $R_1 = R \mod PS(-1,1,0)$, $R_1 \geq \eta > 0$, positive self adjoint operator on $L^2(T^n)$. That we can do this follows from Gårding's inequality. Hence

$$\begin{aligned}\frac{d}{dt}(R_1u,u) &= (R_1'u,u) + (R_1u',u) + (R_1u,u') \\ &= (R_1Ku + f, u) + (R_1u, Ku + f) + (R_1'u,u) \\ &= ((R_1K + K^*R_1)u,u) + (f,u) + (R_1u,f) + (R_1'u,u) \\ &\leq C\|u\|^2 + C\|f\|^2,\end{aligned}$$

since property (ii) of the symmetrizer yields $R_1K + K^*R_1 \in PS(0,1,0)$.

Gronwall's inequality applied to this yields again our energy inequality $\| u(t) \|^2 \leq C \| u(0) \|^2 + C \int_0^t \| f(\tau) \|^2 d\tau$, $0 \leq t \leq T$, and similarly,

for all real S , $\| u(t) \|_S^2 \leq C_S \| u(0) \|_S^2 + C_S \int_0^t \| f(\tau) \|_S^2 d\tau$, $0 \leq t \leq T$, valid for any $u \in C^1([0, \tau], H^{S+1}(\mathbb{T}^n))$, if $u' = Ku + f$.

From this energy inequality, one derives, in exactly the same fashion as in the previous section, the following existence and uniqueness theorem.

<u>THEOREM 1 :</u> If $\frac{\partial}{\partial t} - K$ is strictly hyperbolic, then the Cauchy problem

(1)
$$\frac{\partial}{\partial t} u = Ku + f$$
$$u(0) = \varphi_0$$

has a unique solution $u \in C([0, \tau], H^S(\mathbb{T}^n))$, given $\varphi_0 \in H^S(\mathbb{T}^n)$, $f \in C([0, \tau], H^S(\mathbb{T}^n))$.

This next theorem follows from the reduction technique of section 1.

<u>THEOREM 2 :</u> If L is a strictly hyperbolic operator of order m, then the Cauchy problem

(2)
$$L u = f$$
$$u(0) = \varphi_0$$
$$\frac{\partial}{\partial t} u(0) = \varphi_1$$
$$\vdots$$
$$\frac{\partial^{m-1}}{\partial t^{m-1}} u(0) = \varphi_{m-1}$$

has a unique solution $u \in C([0,\tau], H^S(\mathbb{T}^n))$, given $\varphi_j \in H^{S-j}(\mathbb{T}^n)$, $f \in C([0,\tau], H^{S-m+1}(\mathbb{T}^n))$.

Exercise 2: Let $u \in \mathcal{D}'(\mathbb{R} \times \mathbb{T}^n)$ and suppose u satisfies the equation $(\frac{\partial}{\partial t} - K)u = 0$ where $\frac{\partial}{\partial t} - K$ is symmetric or strictly hyperbolic. If $u = 0$ for $t < 0$, prove that $u = 0$.

Exercise 3: Let L be a strictly hyperbolic operator with smooth coefficients on $\mathbb{R} \times \mathbb{R}^n$. By altering the coefficients of L, and the initial data $f, \varphi_0, \cdots, \varphi_{m-1}$ in the x variables to be periodic outside some large subset of $\mathbb{R}^n$, with large period, and then applying Theorem 3, deduce the following local existence theorem: Given any relatively compact open set $(-a,a) \times U \subset \mathbb{R} \times \mathbb{R}^n$, there is a $u \in C(\mathbb{R}, H^S(\mathbb{R}^n))$ such that $\frac{\partial^j}{\partial t^j} u(0) = \varphi_j$, $j = 0, \cdots, m-1$, and such that, on $(-a,a) \times U$, $Lu = f$, given $\varphi_j \in H^{S-j}(\mathbb{R}^n)$ and $f \in C(\mathbb{R}, H^{S-m+1}(\mathbb{R}^n))$.

Exercise 4: Discuss the Cauchy problem for a differential operator $L = \frac{\partial^m}{\partial t^m} - \sum_{j=0}^{m-1} A_{m-j}(x,t,D_x) \frac{\partial^j}{\partial t^j}$ where $A_{m-j}(x,t,D_x)$ are themselves $k \times k$ systems of differential operators.

Finally, we shall remark that often one wants to handle a Cauchy problem for a strictly hyperbolic equation $Lu = f$, where the initial data are prescribed, not on a hypersurface $t = t_0$, but on a curved surface

S in $\mathbb{R} \times \mathbb{R}^n$. This can be converted into the standard sort of initial value problem by a change of coordinates, under a certain condition on S, which we now describe.

If $L(x,t,D)$ is strictly hyperbolic with respect to the hyperplanes $t = \text{const.}$, with symbol $L(x,t,\xi,\tau)$, write $\omega = (\xi, \tau)$, $N = (0,1)$, the unit normal to the planes $t = \text{const.}$ The condition that L is strictly hyperbolic can be rewritten as follows: for all $\omega \in \mathbb{R}^{n+1}$, not proportional to N, all (x_0, t_0), the polynomial equation

$$L_m(x_0, t_0, \omega + \tau N) = 0$$

has m distinct real roots $\tau_1, \cdots, \tau_m$.

Definition: A vector $\nu = (\nu', \nu_0)$ at (x_0, t_0) is said to be timelike if $\frac{|\nu_0|}{|\nu'|} > \frac{|\tau|}{|\xi|}$ for all (τ, ξ) with $L_m(x_0, t_0, \xi, \tau) = 0$, $\xi \neq 0$. A smooth surfact S in $\mathbb{R}^{n+1}$ is said to be space like for L at (x_0, t_0) if each normal vector to S is timelike.

Proposition: If S is space like for the strictly hyperbolic operator L, and if ν is the normal to S, then the equation

$$L(x_0, t_0, \omega + \tau\nu) = 0$$

has m distinct real roots $\tau_1, \cdots, \tau_m$, if ω is not proportional to ν. Also, S is noncharacteristic for L.

The proof is fairly straightforward and will be left to the reader.

From this we have the following important fact. If S is a space like surface, one can construct a smooth real valued function $t^1(x, t)$ such that S is a level surface for t^1, and such that every level surface of t^1 is space like for L. Furthermore, if L is expressed in terms of the new coordinates (x, t^1), L is strictly hyperbolic with respect to the hypersurfaces $t^1 = \text{const.}$ Thus the Cauchy problem $Lu = f$ can be solved, locally, with Cauchy data given on S.

§4. Finite propagation speed; finite domain of dependence.

Let $L = L(x, t, D_x, D_t)$ be a differential operator of order m on $\mathbb{R}^{n+1}$, with smooth coefficients, which is strictly hyperbolic with respect to the hyperplanes $t = \text{const.}$ Suppose G is a bounded open subset of $\{(x, t) \in \mathbb{R}^{n+1} : t > 0\}$ whose boundary ∂G consists of two parts, $S_1 = \partial G \cap \mathbb{R}^n$ and S_2, a space like surface. The principal result of this section is the following uniqueness theorem.

THEOREM 1: Let $u \in C(\mathbb{R}, H^S(\mathbb{R}^n))$, and suppose that $u(0) = 0$ on S_1 while $Lu = 0$ on G. Then u vanishes on G.

Proof: What we will show is that $\langle u, \varphi \rangle = 0$ for all $\varphi \in C_0^\infty(G)$. To do this, we note that S_1 is also a space like surface for L^*, the adjoint of L, since in fact, the principal symbols of these two operators are essentially identical. (This isn't true for general differential operators the reader should verify that it is true for hyperbolic operators. Hint: it's true for operators with real coefficients.) If $\tilde{S}_2$ denotes the image of S_2 translated in the negative t-direction just a bit, S_2 is also space like. Now let Φ be a local solution to the Cauchy problem

$$L^*\Phi = \varphi$$

$$\Phi|_{\tilde{S}_2} = \varphi$$

$$\vdots$$

$$\frac{\partial^{m-1}}{\partial \nu^{m-1}} \Phi|_{\tilde{S}_2} = 0$$

We can continue Φ to be zero above $\tilde{S}_2$, so Φ is a smooth function satisfying $L^*\Phi = \varphi$, vanishing in a nbd of the closure of S_2. Thus we can write $0 = \langle\Phi, Lu\rangle = \langle L^*\Phi, u\rangle = \langle\varphi, u\rangle$, and we are done. The integration by parts is justified since either Φ or u, together with all Cauchy data, vanish on $\partial\Omega$, Φ vanishing in a nbd of the corner of this boundary.

This type of argument was first used by Holmgren to prove uniqueness in the Cauchy problem for differential operators with analytic coefficients, and non-characteristic boundary. See [13] or [31] ; in that case, the existence theorem used was the Cauchy-Kowalevski theorem.

Definition: Let $(s_0, t_0) \in \mathbb{R}^{n+1}$. If there is a bounded domain G with $(x_0, t_0) \in G$, whose boundary consists of a portion S_1 of $\mathbb{R}^n = \{(x,t): t=0\}$ and a space-like surface S_2 for L, we say S_1 is a domain of dependence for (x_0, t_0), and we say that (x_0, t_0) has a finite domain of dependence.

The following theorem is a simple consequence of Theorem 1 together with the results of the previous section.

THEOREM 2: Let L be a differential operator of order m, strictly hyperbolic with respect to the surface $t = \text{const.}$

(a) If the vector (ν', ν_0) is timelike whenever $\frac{|\nu_0|}{|\nu'|} > M$, then any $(x_0, t_0) \in \mathbb{R}^{n+1}$ has the bounded domain of dependence $B_{M|t_0|}(x_0) = \{x \in \mathbb{R}^n : |x - x_0| \leq M|t_0|\}$.

(b) Given $\varphi_j \in \mathcal{D}'(\mathbb{R}^n)$, $f \in C(\mathbb{R}, \mathcal{D}'(\mathbb{R}^n))$, the Cauchy problem

$$\begin{aligned} Lu &= f \\ \varphi(0) &= \varphi_0 \\ &\vdots \\ \frac{\partial^{m-1}}{\partial t^{m-1}} \varphi(0) &= \varphi_{m-1} \end{aligned}$$

has a unique solution $u \in C(\mathbb{R}, \mathcal{D}'(\mathbb{R}^n))$, if each point (x_0, t_0) of $\mathbb{R}^{n+1}$ has a finite domain of dependence.

(c) If the conditions of (a) are satisfied, and if supp $\varphi_j \subset K$, $f = 0$, then the solution of the above Cauchy problem has the property that, for each $t \in \mathbb{R}$, supp $u(\cdot, t)$ $K_{M|t|}$, where $K_r = \{x \in \mathbb{R}^n : \text{dist}(x, K) \leq r\}$.

The last property is known as finite speed of propagation of a signal.

Exercise 5: Suppose $L = \dfrac{\partial^m}{\partial t^m} + \sum_{j=0}^{m-1} A_{m-j}(x, t, D_x) \dfrac{\partial^j}{\partial t_j}$ is a differential operator of order m, strictly hyperbolic with respect to the surfaces $t = \text{const.}$, and suppose every $(x_0, t_0) \in \mathbb{R}^{n+1}$ has a finite domain of dependence. Then there is a unique $R \in C(\mathbb{R}, H_{loc}^{n/2 - \varepsilon + m-1}(\mathbb{R}^n))$ such that

$$\begin{aligned} LR &= 0 \\ R(0) &= 0 \\ &\vdots \\ \frac{\partial^{m-2}}{\partial t^{m-2}} R(0) &= 0 \\ \frac{\partial^{m-1}}{\partial t^{m-1}} R(0) &= \delta \quad . \end{aligned}$$

Let $\Phi \in \mathcal{D}'(\mathbb{R}^{n+1})$ be defined by $\Phi(t) = R(t)$ for $t > 0$, $\Phi(t) = 0$ for $t < 0$. Prove that $L\Phi = \delta$. What can you say about the support of Φ ?

Exercise 6: Concoct an example of a strictly hyperbolic operator L whose coefficients near infinity behave so badly that not all points $(x_0, t_0) \in \mathbb{R}^{n+1}$ have a finite domain of dependence.

§5. The vibrating membrane problem.

In this section we shall merely indicate how to obtain a hyperbolic equation for one physical process. For a further introduction to the equations governing continua, see Goldstein [2].

We suppose we have a thin membrane, pulled tight, which vibrates slightly in the direction perpendicular to the plane in which it is to lie; call it the x-y plane. If $\varphi(x, y)$ is the distance above or below that plane, then for φ small, Hook's law says the potential energy in the membrane, due to the slight displacement it has undergone, is approximately

$$V = \int_\Omega k|\nabla\varphi|^2 \, dV$$

where Ω is the plane occupied by the membrane and k is an appropriate factor, $k > 0$ on Ω. On the other hand, the kinetic energy due to this vibration is

$$T = \int_\Omega \rho |\varphi_t|^2 \, dV$$

where ρ is half the mass density. Now Hamilton's principle says that the action $\int_{t_0}^{t_1} (T-V)\, dt$ is stationary. That is, we are led to the calculus of variations problem

$$\frac{d}{d\eta} I(\varphi + \eta\psi)\Big|_{\eta=0} = 0$$

for all $\psi \in C_0^\infty(\Omega \times (t_0, t_1))$ where $I(\varphi) = \int_{t_0}^{t_1} \int_\Omega \{k|\nabla\varphi|^2 - \rho|\varphi_t|^2\} \, dVdt$.

Hence we get, for all $\psi \in C_0^\infty(\Omega \times \mathbb{R})$,

$$0 = \frac{d}{d\eta} I(\varphi + \eta\psi)\Big|_{\eta=0}$$

$$= \int_{t_0}^{t_1} \int_\Omega \{2k(\nabla\varphi, \nabla\psi) - 2\rho\,\varphi_t\,\psi_t\}\, dVdt$$

$$= -2\int_{t_0}^{t_1} \int_\Omega \left(\sum_{j=1}^{2} \frac{\partial}{\partial x_j}\left(k\frac{\partial}{\partial x_j}\right) - \frac{\partial}{\partial t}(\rho\varphi_t)\right)\psi\, dVdt \quad .$$

Since this integral vanishes for all $\psi \in C_0^\infty(\Omega \times (t_0, t_1))$, we must have

$$\text{(1)} \qquad \sum_{j=1}^{2} \frac{\partial}{\partial x_j}\left(k\frac{\partial}{\partial x_j}\varphi\right) - \frac{\partial}{\partial t}\left(\rho\frac{\partial}{\partial t}\varphi\right) = 0 \quad .$$

The reader should verify that this is a second order, strictly hyperbolic equation, if $\rho, k > C > 0$ on Ω.

For such a problem, we must specify boundary data of φ as well as its initial data. For instance, if our membrane is securely fastened to the head of a drum, we have $\varphi = 0$ on $\partial\Omega$, for all t. In these notes, we have not touched on such mixed initial-boundary value problems. This particular mixed problem can be handled, in a spirit similar to preceding sections, using the fact that the total energy

$$E = T + V = \int_\Omega [\rho|\varphi_t|^2 + k|\nabla\varphi|^2]\, dV$$

is constant.

This fact, that energy is conserved in physical systems, or doesn't increase in dissipative systems, or doesn't increase too fast when there is an input of energy, etc., is the origin for the type of inequalities which we have called energy inequalities in previous sections.

Exercise : If φ satisfies (1), $\varphi = 0$ in $\partial\Omega$, prove that $\frac{dE}{dt} = 0$.

§ 6. Parabolic evolution equations.

Definition: If $-K \in PS(S, 1, 0)$ is strongly elliptic of order S, on M, then the operator $\frac{\partial}{\partial t} - K$ is a strongly parabolic operator on $\mathbb{R} \times M$.

Here we take $K = K(t)$ to be a smooth one-parameter family of operators on M, as before. We take M to be any compact C^{∞} manifold. If $(\, , \,)$ is an L^2 inner product on M, then $(\frac{\partial}{\partial t} - K) u = f$ yields

$$\frac{d}{dt}(u, u) = (Ku + f, u) + (u, Ku + f)$$

$$= ((K + K^*)u, u) + 2 \operatorname{Re}(u, f)$$

$$\leq C\|u\|^2 + C\|f\|^2$$

where Gårding's inequality is applied to the strongly elliptic operator $-K$ to give this inequality. Thus

$$\|u(t)\|^2 \leq C\|u(0)\|^2 + C\int_0^t \|f(\tau)\|^2 \, d\tau \qquad 0 \leq t \leq T$$

for $u \in C^1([0,T], H^S(M))$. From this energy inequality, existence and uniqueness of solutions to the initial value problem follows, by the same sort of argument as given in Section 2.

Note that, unlike hyperbolic equations, strongly parabolic equations can only be solved for positive t, not for negative t.

To solve the initial value problem for $\frac{\partial}{\partial t} - K$, for $K \in PS(S, 1, 0)$, we only need $-\operatorname{Re} \tau(x, t, \xi) \geq C|\xi|^S$ for any eigenvalue τ of $\sigma_K(x, t, \xi)$, $|\xi| \geq 1$. To show this involves constructing a symmetrizer, in the spirit of section 3. We exhibit a more general construction in the next chapter.

Solutions to parabolic equations enjoy additional smoothness conditions, which we deal with in a more general context in the next chapter.

The next few exercises give another approach to parabolic equations in the temporally homogeneous case.

Exercises

8. Suppose $A \in PS(m, 1.0)$ on a compact manifold M, is such that $-A$ is strongly elliptic of order m. Prove that the spectrum and numerical range of $A - \lambda_0$ is a cone in the negative half plane in $\mathbb{C}$ if λ_0 is sufficiently large. Deduce that A generates a strongly continuous semigroup e^{tA}; in fact, A generates a holomorphic semigroup. (Hint: look at the argument on p. 102 of Agmon [1].)

9. If $\varphi_0 \in L^2(M)$, $e^{tA} \varphi_0 \in C^\infty(m)$ for each $t > 0$. If $u(\cdot, t) = e^{tA} \varphi_0$, then $u \in C^\infty((0, \infty) \times M)$.

Hint: $e^{tA} \varphi_0 \in \mathcal{D}(A^k)$ for all k since e^{tA} is a holomorphic semigroup.

10. Prove that e^{tA} is a strongly continuous semigroup of operators on each Sobolev space $H^S(M)$, $S \in \mathbb{R}$.

11. Let A generate a holomorphic semigroup on a Hilbert space H. Prove that for all $\varphi \in H$,

$$\int_0^1 \mathrm{Re}\,(-Ae^{tA}\varphi, e^{tA}\varphi)\,dt = \frac{1}{2}\|\varphi\|^2 - \frac{1}{2}\|e^A\varphi\|^2 .$$

Deduce that, when A is as in exercises 8-10, $\varphi \in H^S(M)$, $u = e^{tA}\varphi$, then $u \in L^2([0,1], H^{S+\frac{m}{2}}(M))$.

12. In the situation of exercise 11, if $m = 1$, show that $\varphi \in H^S(M)$ implies $u \in H^{S+\frac{1}{2}}([0,1] \times M)$.

Hint: See Theorem 4.3.1 of Hörmander [31].

§7. References to further work.

One big subject we have not touched on is mixed problems for hyperbolic equations. For symmetric hyperbolic equations, see [13], [52], and for another general class, see Kreiss [46] and Rauch [68].

A topic which is interesting both for mixed problems and the pure initial value problem is scattering theory; see [52], [53]. Closely allied to this is the method of geometrical optics. Using this construction, Lax [49], constructed a very accurate approximate solution to the Cauchy problem. This method has been extended and generalized by Hörmander; see [39], [18], [38]. One application of these methods is the determination of singularities of solutions to hyperbolic (and more general) equations. Using the theory of pseudo-differential operators, we shall obtain such results in Chapter VI.

The problem of how singularities of solutions to boundary value problems propagate is still very much open, especially when there are tangential bicharacteristics. See, however, Nirenberg [91].

Another important topic we have not touched on is difference schemes, approximating solutions to initial value problems. See [70]. A calculus of pseudo-difference operators has been developed to treat problems in this area. See Yumuguti and Nogi [84] and Lax and Nirenberg [51], and also Vaillancourt [82].

Certain mixed problems for second order hyperbolic equations can be studied by means of functional analysis and the elliptic theory. One approach, due to Phillips, is outlined in the beginning of [69].

Another use to which energy inequalities via pseudo-differential operators have been put, first by Calderon, is the study of uniqueness in the Cauchy problem, a topic we have been unable to touch on in these lectures. See Kumano-Go [47] , or Nirenberg [91].

Chapter V. Elliptic Boundary Value Problems; Petrowsky Parabolic Operators

Let M be a smooth compact manifold with boundary, ∂M. Suppose L is an elliptic operator of order m on M, and B_j, $j = 1,\ldots,\nu$ are differential operators of order $m_j \leq m-1$ defined in a neighborhood of ∂M. We want to investigate the boundary value problem.

$$(1) \qquad Lu = f$$

$$B_j u|_{\partial M} = g_j \quad 1 \leq j \leq \nu$$

Two properties we shall investigate are existence of solutions and smoothness. As we shall see in section 2, existence will follow from the a priori estimates needed to prove smoothness results, via functional analysis.

To prove smoothness results for the solution u of (1), we only need to worry about how u behaves near the boundary, since Chapter III gives us the desired smoothness properties of u on the interior of M.

We proceed to reduce the problem near the boundary to a standard form. First, the collar neighborhood theorem says there is a neighborhood of ∂M in M diffeomorphic to $I \times \partial M$, where $I = [0,1]$. For a proof of this, see Milnor's Topology from the Differentiable Viewpoint. If we write these coordinates as (t,x), L takes the form

$$L = \frac{\partial^m}{\partial t^m} + \sum_{j=0}^{m-1} A_j(x,t,D_x) \frac{\partial^j}{\partial t^j}$$

where $A_j(x,t,D_x)$ is a smooth one parameter family of differential operators on ∂M, of order $m-j$. This can be reduced to a first order system $\frac{\partial}{\partial t} + K\Lambda$, $K \in PS(0,1,0;\ M)$, by the same technique as used in Chapter IV.

Thus, with $u_j = \left(\frac{\partial}{\partial t}\right)^{j-1} \Lambda^{m-j}\, u$, and if

$$B_j = \sum_{k=0}^{m_j} b_k^j\,(x,D_x) \frac{\partial^k}{\partial t^k}\,,\quad b_k^j \text{ of order } m_j - k,$$

the boundary value problem (1) becomes

$$\frac{\partial}{\partial t}\begin{pmatrix} u_1 \\ \vdots \\ u_m \end{pmatrix} = -K\Lambda \begin{pmatrix} u_1 \\ \vdots \\ u_m \end{pmatrix} + \begin{pmatrix} 0 \\ \vdots \\ 0 \\ f \end{pmatrix}$$

$$\Lambda^{m-m_j-1} \sum_{k=0}^{M_j} b_k^j\,(x,D_x)\,\Lambda^{k+1-m}\, u_{k+1}\,(0) = \Lambda^{m-m_j-1}\, g_j$$

Here $-K\Lambda$ is the first order operator denoted by K on p. 59. We now discuss such systems, in greater generality.

Let $K = K(t)$ be a smooth one parameter family of operators in $PS(0,1,0)$ on ∂M, and let $\Lambda^s \in PS(s,1,0)$ be any fixed positive definite elliptic operator on ∂M of order $s > 0$.

Def: The operator $H = \frac{\partial}{\partial t} + K\Lambda^s$ on $[0,1] \times \partial M$ is said to be Petrowsky parabolic if $K = K_0 + K_1$ where $K_1 \in PS(-1,1,0)$ and $\sigma_{K_0}(x,t,\xi)$ is homogeneous of degree zero in ξ for $|\xi| \geq 1$, and if the k x k matrix $\sigma_{K_0}(x,t,\xi)$ has <u>no pure imaginary eigenvalues</u> for $(t,x) \in I \times \partial M$, $|\xi| \geq 1$.

The case $s = 1$ is the elliptic case. Reducing an elliptic operator L to a first order system leads to such a system. We deal with the more general parabolic case, because it requires no additional work. The approach we take here follows Polking [66]. The reduction to the boundary we make was first conceived by Calderon, for general regular elliptic problems, and carried out by Hörmander [33] for the subelliptic case as well. See also Seeley [73].

We now construct some pseudo differential operators which we will need in section 1 to prove a priori estimates.

If $t \in I$ is fixed, define $\sigma_{E(t)}(x,\xi)$ to be the projection onto the direct sum of the positive generalized eigenspaces of $\sigma_{K_0(t)}(x,\xi)$, defined by $\sigma_{E(t)}(x,\xi) = \frac{1}{2\pi i}\int_\gamma (\zeta - \sigma_{K_0(t)}(x,\xi))^{-1}\, d\zeta$, where γ is a curve in the right half plane containing the positive eigenvalues of $\sigma_{K_0(t)}(x,\xi)$. Clearly $\sigma_{E(t)}(x,\xi)$ is homogeneous of degree zero in ξ, for $|\xi| \geq 1$, so is the principal symbol of a smooth one parameter family of operators $E = E(t) \in PS(0,1,0;\ \partial M)$. Essentially, applying $E(t)$ separates the equation $Hu = f$ into a forward and a backward evolution equation. The reader might verify that we can assume $E(t)^2 = E(t)$, for all t, but we don't need this.

If $A = (2E - 1) K$, then $\sigma_A(x,t,\xi)$ has only positive eigenvalues, for large ξ. The next task for us is to construct a smooth one parameter family $P = P(t)$ of pseudo differential operators on ∂M such that PA is strongly elliptic. This plays the role that the "symmetrizer" played in Chapter IV.

Lemma: Let A be a $k \times k$ matrix and suppose the eigenvalues of A all have positive real part. Then there is a positive matrix P such that $PA + A^*P$ is also positive.

Proof: For any $\varepsilon > 0$, A is similar to $D_0 + \varepsilon D_1$ where D_0 is diagonal and $D_0 + D_1$ in the Jordan canonical form of A. If $QAQ^{-1} = D_0 + \varepsilon D_1$, write $Q = UP_1$ with U unitary and P_1 positive. Then $P_1AP_1^{-1} = U^*D_0U + \varepsilon U^*D_1U = N + B_\varepsilon$ where N is normal and $\|B_\varepsilon\| \leq \varepsilon$. The eigenvalues of N and A are the same, so there is an $\eta > 0$, independent of ε, such that $N + N^* \geq \eta$. Taking $\varepsilon < \frac{1}{3}\eta$, the matrix $N + B_\varepsilon$ is accretive. If $P = P_1^{\,2}$, it follows that $PA + A^*P = P_1(N + B_\varepsilon) P_1 + P_1(N^* + B_\varepsilon^*) P_1$ is positive.

Before we proceed, two observations are in order. First, for a given positive matrix P, the set of A such that $PA + A^*P$ is positive, is open. Second, for a given compact set K of $k \times k$ matrices, $\{P : P > 0,\ PA + A^*P > 0$ for all $A \in K\}$ is an open, convex set of matrices.

Proposition 1: If $A = (2E - 1)\ K$ is the smooth one parameter family of operators defined above, so that the eigenvalues of the principal symbol $a(t,x,\xi)$ of A have positive real part, for $|\xi| \geq 1$, there is a smooth one parameter family $P = P(t) \in PS(0,1,0)$ such that σ_P is positive definite and PA is strongly elliptic.

Proof: With the above lemma and subsequent observation, it is easy using a partition of unity to construct σ_P on the cosphere bundle of ∂M such that σ_P and $\sigma_P\sigma_A + \sigma_A{}^*\sigma_P$ are positive. Extend $\sigma_P(x,\xi)$ to be homogeneous of degree zero in ξ, and let P be a smooth family of pseudo differential operators with σ_P as principal symbol. The proof is complete.

Finally, let us define the spaces of functions which we will need. Def: $H_{<k,r>} = \{u \in H^{-\infty}(I \times \partial M) : (\frac{\partial}{\partial t})^j\ \Lambda^{s(k-j)+r} u \in L^2(I \times \partial M);\ 0 \leq j \leq k\}$. The norm on $H_{<k,r>}$ is defined by

$$\|u\|^2_{<k,r>} = \sum_{j=0}^{k} \left\|\left(\frac{\partial}{\partial t}\right)^j \Lambda^{s(k-j)+r} u\right\|^2_{L^2(I \times \partial M)}$$

Note that the case $s = 1$ yields the familiar space $H_{(k,s)}$

of [31], with $\|u\|^2_{(k,r)} = \sum_{j=0}^{k} \left\| \left(\frac{\partial}{\partial t}\right)^j \Lambda^{k-j+r} u \right\|^2$

We shall denote the norm on $H^s(\partial M)$ by $|\ |_s$.

The following proposition is a generalization of the standard result for Sobelov spaces, given in [55]. The reader may try to prove it himself, or see Polking [66].

Proposition 2: The map $u \to \left(\frac{\partial}{\partial t}\right)^\ell u|_{\partial M}$ gives rise to a continuous map of $H_{<k,r>}$ onto $H^{s(k-\ell-1/2)+r}(\partial M)$.

1. A priori estimates and regularity theorems.

Acting on the idea that the projection E splits $K\Lambda^s$ up into a forward and backward evolution operator, we will differentiate $(P\Lambda^{s/2} Eu, \Lambda^{s/2} Eu)$ and $(P\Lambda^{s/2}(1-E)u, \Lambda^{s/2}(1-E)u)$ with respect to t, in the spirit of the energy inequalities of the last chapter, and add the two up. Some new features, naturally, will appear. Let $H = \frac{\partial}{\partial t} + K\Lambda^s$.

$$\frac{d}{dt}(P\Lambda^{s/2} Eu, \Lambda^{s/2} Eu)$$

$$= (P\Lambda^{s/2} Eu', \Lambda^{s/2} EU) + (P\Lambda^{s/2} Eu, \Lambda^{s/2} Eu') + (P'\Lambda^{s/2} Eu, \Lambda^{s/2} Eu) + \ldots$$

$$= (P\Lambda^{s/2} E(Hu - K\Lambda^s u), \Lambda^{s/2} Eu) + (P\Lambda^{s/2} Eu, \Lambda^{s/2} E(Hu - K\Lambda^s u)) + \ldots$$

$$= 2\,\mathrm{Re}\,(P\,E\,Hu, \Lambda^s Eu) - 2\,\mathrm{Re}\,(PEK\Lambda^s Eu, \Lambda^s Eu) + \ldots$$

1) $$= 2\,\mathrm{Re}\,(P\,E\,Hu, \Lambda^s Eu) - ([PA+A^*P]\,\Lambda^s Eu, \Lambda^s Eu) + R_1(u)$$

We used $E \equiv (2E-1)\,E \bmod PS(-1,1,0)$ to conclude that $PEK\Lambda^s E \equiv PA\Lambda^s E \bmod PS(s-1,1,0)$. Here and below, $R_j(u)$ denotes a remainder term with the property that

$$|R_j(u)| \leq C|u|_s\,|u|_{s-1}.$$

Similarly, we have

) $$\frac{d}{dt}(P\Lambda^{s/2}(1-E)u, \Lambda^{s/2}(1-E)u)$$

$$= 2\,\mathrm{Re}\,(P(1-E)\,Hu, \Lambda^s(1-E)u) + ([PA+A^*P]\,\Lambda^s(1-E)u, \Lambda^s(1-E)u)$$

$$+ R_2(u)$$

Adding (1) and (2), and applying Gårding's inequality to $PA + A^*P$, we get

$$C|\Lambda^s u|^2 \le C|\Lambda^s E u|^2 + C|\Lambda^s(1-E) u|^2$$

$$\le C([PA + A^*P] \Lambda^s Eu, \Lambda^s Eu) + ([PA + A^*P] \Lambda^s (1-E) u, (1-E)u) + R_4(u)$$

$$\le \frac{d}{dt} (P\Lambda^{s/2} (1-E) u, \Lambda^{s/2}(1-E)u) - \frac{d}{dt}(P\Lambda^{s/2} Eu, \Lambda^{s/2} Eu) + C|Hu| |\Lambda^s u| + R_5(u)$$

Integrating with respect to t, we get

$$(3) \qquad \|u\|^2_{(0,s)} + |(1-E(0))\, u(0)|^2_{s/2} + |E(1)\, u(1)|^2_{s/2} \le C\|Hu\|^2_{(0,0)} + C|E(0)\, u(0)|_{s/2} + C|(1-E(1))\, u(1)|^2_{s/2} + C\|u\|^2_{(0,0)}$$

Since $\|u\|^2_{<1,0>} \le C\|u'\|^2_{(0,0)} + C\|u\|^2_{(0,s)} \le C\|Hu\|^2_{(0,0)} + C\|u\|^2_{(0,s)}$, we get

$$(4) \qquad \|u\|^2_{<1,0>} \le \|Hu\|^2_{(0,0)} + C|E(0)u(0)|^2_{s/2} + C|(1-E(1))\, u(1)|^2_{s/2}$$

From now on, let $R(t) = t + (1-2t)\, E(t)$, and let $|u|^2_\tau = |u(1)|^2_\tau + |u(0)|^2_\tau$. The next higher order a priori estimate we derive by substituting $\Lambda^t u$ for u in (4).

$$\|u\|^2_{<1,t>} = \|\Lambda^t u\|^2_{<1,0>}$$

$$\leq \|H\Lambda^t u\|^2_{(0,0)} + C|R\Lambda^t u|^2_{s/2} + C\|\Lambda^t u\|^2_{(0,0)}$$

$$\leq C\|Hu\|^2_{(0,t)} + C|Ru|^2_{s/2+t} + C\|u\|^2_{(0,s+t-1)}$$

$$+ C|u|^2_{s/2+t-1} .$$

Now $|u|^2_{s/2+t-1} \leq C\|u\|^2_{<1,t-1>} \leq C\|Hu\|^2_{(0,t-1)} + C\|u\|^2_{(0,s+t-1)}$, so

(5) $$\|u\|_{<1,t>} \leq C\|Hu\|^2_{(0,t)} + C|Ru|^2_{s/2+t} + C\|u\|^2_{(0,0)} .$$

Similarly we obtain $|(1-R)u|^2_{s/2+t} \leq C\|Hu\|^2_{(0,t)} + C|Ru|^2_{s/2+t} + C\|u\|^2_{(0,0)}$, and substituting $(1-R)u$ for u in this inequality, using the fact that the principal symbols of $R(t)$ and $K(t)$ commute, for each t, we obtain the important inequality

(6) $$|(1-R)u|^2_{s/2+t} \leq C\|Hu\|^2_{(0,t)} + C\|u\|^2_{(0,s+t-1)} + C|u|^2_{s/2+t-1}.$$

Now suppose we have a boundary operator $B \in PS(0,1,0,\partial M)$, and suppose the following a priori inequality is valid (for each τ)

*) $$|g|^2_{\tau-\delta} \leq C|(1-R)g|^2_{\tau} + C|Bg|^2_{\tau} + C|g|^2_{\tau-1} \qquad g \in C^\infty(\partial M)$$

for some δ with $0 \leq \delta < 1$. Inequality (6) yields

$$|u|^2_{s/2+t-\delta} \leq C\,\|Hu\|^2_{(0,t)} + C\,\|u\|^2_{(0,s+t-1)} + C|u|^2_{s/2+t-1}$$

$$+ C|Bu|^2_{s/2+t}.$$

$$\therefore \|u\|^2_{<1,t-\delta>} \leq C\,\|Hu\|^2_{(0,t-\delta)} + C|Ru|^2_{s/2+t-\delta} + C\,\|u\|^2_{(0,0)}$$

$$\leq C\,\|Hu\|^2_{(0,t-\delta)} + C|u|^2_{s/2+t-\delta} + C\,\|u\|^2_{(0,0)}$$

(7) $$\therefore \|u\|^2_{<1,t-\delta>} \leq C\,\|Hu\|^2_{(0,t)} + C|Bu|^2_{s/2+t} + C\,\|u\|^2_{(0,0)}$$

since we can eliminate the term $|u|^2_{s/2+t-1}$ as we did in deriving (5).

Our main a priori estimate is the following.

Theorem 1: Suppose (*) is satisfied. Then for any integer $k \geq 1$,

(8) $$\|u\|^2_{<k,t-\delta>} \leq C\|Hu\|^2_{<k-1,t>} + C|Bu|^2_{s/2+t+(k-1)s} + C\,\|u\|^2_{(0,0)}$$

for all $u \in H_{<k,t>}(I \times \partial M)$.

Proof: Proceed by induction, the case $k = 1$ being inequality (7). If (8) is valid for k, we get

$$\|u\|^2_{<k+1,t-\delta>} \le C\,\|u'\|^2_{<k,t-\delta>} + C\,\|u\|^2_{<k,t-\delta+s>}$$

$$\le C\,\|Hu\|^2_{<k,t-\delta>} + C\,\|u\|^2_{<k,t-\delta+s>}.$$

We estimate the right hand side of the inequality by (8) for k, with t replaced by $t+s$, and the induction step follows.

To derive the regularity theorem corresponding to the estimate (8), we make use of a Friedrichs' mollifier with respect to the variable $x \in \partial M$, J_ε.

Lemma: if $u \in H_{<1,r-1>}$ and $H u \in H_{(0,r)}$, $Bu \in H^{r+s/2}(\partial M)$, then $u \in H_{<1,r-\delta>}$, provided (*) holds.

Proof: For each $\varepsilon > 0$, $J_\varepsilon u \in H_{<1,+\infty>}$, so (8) applies, and we get

$$\|J_\varepsilon u\|^2_{<1,r-\delta>} \le C\,\|HJ_\varepsilon u\|^2_{<u,r>} + C|BJ_\varepsilon u|^2_{s/2+r} + C\,\|J_\varepsilon u\|^2_{(0,0)}$$

$$\le C\,\|Hu\|^2_{<0,r>} + C|Bu|^2_{s/2+r} + C\,\|u\|^2_{(0,r+s-1)}$$

$$+ C|u|^2_{s/2+r-1},$$

using the fact that $[J_\varepsilon, K\Lambda^s]$ is bounded in $PS(s-1,1,0)$ and $[J_\varepsilon,B]$ is bounded in $PS(-1,10)$. Hence

$$\|J_\varepsilon u\|^2_{<1,r-\delta>} \le C\,\|Hu\|^2_{<0,r>} + C|Bu|^2_{s/2+r} + C\,\|u\|^2_{<1,r-1>},$$

a bound which is <u>independent</u> of ε. Thus $\{J_\varepsilon u: 0 < \varepsilon \leq 1\}$ is a bounded subset of $H_{1,r-\delta}$, so some subsequence $J_{\varepsilon_n} u \to \tilde{u}$ as $n \to \infty$, weakly in $H_{<1,r-\delta>}$. But $J_\varepsilon u \to u$ in $H_{<1,r-1>}$, so $u = \tilde{u} \in H_{<1,r-\delta>}$, as desired.

Note: The hypothesis $u \in H_{<1,r-1>}$ can be replaced by $u \in H_{<1,-\infty>}$. Proof by induction.

Proposition: Let $u \in H_{(0,\sigma)}$, $Hu \in H_{(0,r)}$, and $Bu \in H^{r+s/2}(\partial M)$. Then $u \in H_{<1,r-\delta>}$, if (*) holds.

Proof: Since $u \in H_{(0,\sigma)}$, $u' = Hu - K\Lambda^s u \in H_{(0,\sigma')}$ where $\sigma' = \min(r,\sigma-s)$. Thus $u \in H_{<1,\sigma'>}$, and the above note yields our result.

The following is our regularity theorem.

Theorem 2: if $u \in H_{(0,\sigma)}$, $Hu \in H_{<k,t>}$, and $Bu \in H^{s/2+t+ks}(\partial M)$, then $u \in H_{<k+1,t-\delta>}$, provided (*) holds.

Proof: First note that $u \in H_{(0,\sigma)}$, $Hu \in H_{<k,t>}$ implies $u \in H_{<1,t-\delta+ks>}$, by the above proposition. In particular, $Bu|_{\partial M}$ makes sense. Now to deduce $u \in H_{<k+1,t-\delta>}$ it will suffice to infer this from the hypothesis $u \in H_{<k,t-\delta+s>}$; an induction argument finishes off the proof. But $u \in H_{<k,t-\delta+s>} \Rightarrow u' = Hu - K\Lambda^s u \in H_{<k,t-\delta}$ $\Rightarrow u \in H_{<k+1,t-\delta>}$.

Corollary: $u \in H_{(0,\sigma)}(I \times \partial M)$, $Hu \in C^\infty(I \times \partial M)$, $Bu \in C^\infty(\partial M)$ $\Rightarrow u \in C^\infty(I \times \partial M)$, if (*) holds.

Exercise 1: Look up theorem 4.3.1 of Hörmander and then translate theorem 1 and 2 into estimates and regularity theorems for the elliptic boundary value problem $Lu = f$, $B_j u = g_j$. The estimate should look like

$$\|u\|^2_{m+k} \leq \|Lu\|^2_k + C \sum_{j=1}^{\nu} |B_j u|^2_{m+k-m_j-1/2} + C\,\|u\|^2_0 .$$

Exercise 2. Suppose $u \in H^{-k}(I \times \partial M)$ and $Hu = f \in C^\infty(I \times \partial M)$. Prove that $u \in H_{(0,\sigma)}$ for some σ. Hint: $u' = f - K\Lambda^s u \in H^{-k}(I, H^{-k-s}(\partial M))$; $u'' = f' - K\Lambda^s (f - K\Lambda^s u) + K'\Lambda^s u$, etc.

2. Closed range and Fredholm properties.

We begin by proving a couple of functional analytic
results which belong to the standard theory of Fredholm
operators, but which we have not seen in quite the form
we state them in, a form most useful for our purposes.

Proposition 1: Let $T:E \to X$ be a closed linear operator between
Banach spaces, and suppose $K:E \to Y$ is compact. If

$$(1) \quad \|u\|_E \leq C \|Tu\|_X + C \|Ku\|_Y \qquad \forall \ u \ e \ \mathcal{D}(T)$$

then T has closed range.

Proof: Since we can replace E by $\mathcal{D}(T)$ with its graph
topology, we may assume $T:E \to X$ is continuous.

Let $Tu_n \to f$ in X. We need $u \ e \ E$ with $Tu = f$. Let
$V = \ker T$. We divide up the argument into two cases.

(i) If $d(u_n, V) \leq \alpha < \infty$, we can take $v_n \equiv u_n \bmod V$, $\|v_n\| \leq 2\alpha$,

$Tv_n + Tu_n \to f$. Passing to a subsequence, we may assume $Kv_n \to g$
in Y. the inequality (1) yields

$$\|v_m - v_n\| \leq C \|Tv_n - Tv_m\| + C \|Kv_m - Kv_n\| \to 0, \text{ so } v_n \to v$$

and $Tv = f$.

(ii) If $d(u_n, V) \to \infty$, assume $d(u_n, V) > 2$ for all n.

Let $\nu_n \equiv u_n \bmod V$, $d(u_n,V) \leq \|\nu_n\| \leq d(u_n,V) + 1$; $T\nu_n = Tu_n$.

If $w_n = \dfrac{\nu_n}{\|\nu_n\|}$, then $d(w_n,V) \geq \dfrac{d(\nu_n,V)}{d(u_n,V)+1} \geq \dfrac{1}{2}$.

also $\|w_n\| = 1$, so we can assume $Kw_n \to g$ in Y. $Tw_n \to 0$. Thus inequality (1) yields $\|w_n - w_m\| \leq C\,\|Tw_n - Tw_m\| + C\,\|Kw_m - Kw_n\| \to 0$. Thus $w_n \to w$ in E and we see simultaneously that $d(w,V) \geq \frac{1}{2}$ and $Tw = 0$, a contradiction. Hence case (ii) is impossible, and the proof is complete.

Proposition 2: If $T:E \to X$, a closed linear operator between Banach spaces, has closed range, of finite codimension, and if $K:E \to X$ is compact, then $T + K$ has closed range, of finite codimension.

Proof: Without loss of generality, T is continuous. Also, it suffices to prove the proposition assuming T is onto, for if V is a finite dimensional complementary space to R(T), $T \oplus j \;\; E \oplus V \to X$ is onto, and we need only apply such a special case of our proposition to $(T + K) \oplus 0 = (T \oplus j) + (K \oplus (-j))$.

But if T is onto, T* is injective and has closed range. Hence $\|w\|_{X'} \leq C\,\|T^*w\|_{E'}$, $w \in X'$. Hence we have the inequality

$$\|w\|_{X'} \leq C\,\|(T^* + K^*)\,w\|_{E'} + C\,\|K^*w\|_{E'}.$$

From proposition 1 it follows that $(T + K)^*$ has closed range; clearly its kernel is finite dimensional. The proof is now completed by taking adjoints back.

Corollary: Let $T:E \to X$ be a closed linear operator, $j:V \to X$ a compact injection. Suppose that for each $x \in X$, a closed subspace of X of finite codimension, there is a $u \in \mathcal{D}(T)$ such that $Tu - x \in V$. Then T has closed range, of finite codimension.

Proof: The hypothesis says that $T \oplus j : E \oplus V \to X$ has closed range, of finite codimension. Now apply proposition 2.

Exercise 3: Refering to proposition 2, give an example to show that if we only assume that T has closed range, it does not follow that $T + K$ has closed range.

To return to differential equations, we begin by showing that

$$Hu = f$$

$$Ru|_{\partial M} = 0$$

can be solved, if f satisfies a finite number of linear conditions. Thus, let $E = \{u \in H_{<1,t>} : Ru|_{\partial M} = 0\}$ and define $H_0:E \to H_{(0,t)}(I \times \partial M)$ by $H_0 u = Hu$. The a priori estimates of section 1 show that $\ker H_0$ is finite dimensional. (By $Ru|_{\partial M} = 0$ we mean $Ru(0) = 0$ and $Ru(1) = 0$.) We will assume that $R(t)^2 = R(t)$, which can be arranged, for $t = 0$ or 1.

Theorem 1: H_0 is Fredholm.

Proof: We only need show that $R(H_0)$ is closed and of finite codimension. Now $\|u\|^2_{<1,t>} \leq C\,\|H_0 u\|^2_{(0,t)} + C\,\|u\|_{(0,0)}$ for $u \in E$, by the inequality (5) of section 1. By proposition 1, $R(H_0)$ is closed. Now suppose $w \in H_{(0,-t)}$, the dual of $H_{(0,t)}$, and

(2) $\langle H_0 u, w\rangle = 0$ for all $u \in E$. This is true in particular for all $u \in C_0^\infty((0,1) \times \partial M)$, so $H^*w = 0$, where

$$H^* = -\frac{\partial}{\partial t} + \Lambda^s K^*$$

$$\therefore \frac{\partial}{\partial t} w = -H^*w + \Lambda^s K^* w \in H_{(0,-t-s)}, \text{ so } w \in H_{<1,-t-s>}.$$

Hence $w|_{\partial M}$ is well defined and belongs to $H^{-t-s/2}(\partial M)$, and the following identity is true: $0 = \langle H_0 u,w\rangle = \langle Hu,w\rangle - \langle u,H^*w\rangle$ $= (u(1), w(1)) - (u(0), w(0))$ for all $u \in E$, if w satisifies (2).

Now for any $\nu \in H^{r+s/2}(\partial M)$, there is a $u \in H_{<1,r>}$ such that $u|_{\partial M} = (1-R)\nu$. In particular, $Ru|_{\partial M} = 0$, so $u \in E$.

$\therefore ((1-R(0))\nu_0, w(0)) = ((1-R(1))\nu_1, w(1))$ for all ν_0, $\nu_1 \in H^{r+s/2}(\partial M)$, if w satisfies (2). Taking $\nu_0 = 0$, we get $(\nu_1, (1-R(1)^*)\, w(1)) = 0$ for all $\nu_1 \in H^{r+s/2}(\partial M)$, or

$(1-R(1))^* w(1) = 0$. Similarly, $(1-R(0)^*) w(0) = 0$. Then all $w \in H_{(0,-t)}$ satisfying (2) must satisfy the boundary value problem

$$H^*w = 0$$
$$(1-R^*) w|_{\partial M} = 0$$

But $1-R^*$ plays the same role for the Petrowsky parabolic operator H^* as R plays for H, so the set of w satisfying this condition is finite dimensional, which is to say that $R(H_0)$ is finite codimensional.

Next we consider the inhomogeneous boundary value problem

$$Hu = f$$

$$Ru|_{\partial M} = g$$

Given $f \in H_{(0,t)}(I \times \partial M)$, $g \in H^{t+s/2}(\partial M)$, we would like to find a solution $u \in H_{<1,t>}$. Since $R(j)^2 = R(j)$ for $j = 0,1$ we must require that $(1-R)\, g = 0$. Thus let

$H^{t+s/2}_{1-R}(\partial M) = \{g \in H^{t+s/2}(\partial M) : (1-R)\, g = 0\}$. Define

$H_1 : H_{<1,t>} \to H_{(0,t)} \oplus H^{t+s/2}_{1-R}(\partial M)$ by $H_1 u = \{u, Ru|_{\partial M}\}$.

Theorem 2: H_1 is Fredholm.

Proof: Take $F = Kg \in H_{<1,t>}$, $F|_{\partial M} = g$. Consider the equation

$$Hv = f - HF$$

$$Rv|_{\partial M} = 0$$

By theorem 1, there is a $v \in H_{<1,t>}$ solving this equation provided $f - HKg$ satisfies some finite number of linear conditions. In such a case, set $u = v+F$. Then $Hu = f$ and $Ru|_{\partial M} = Ru_{\partial M} + Rg = g$. The Fredholm property of H_1 is now immediate.

Next we consider the boundary value problem

$$(3) \qquad \begin{aligned} Hu &= 0 \\ Bu &= g \end{aligned}$$

where $B: H^{t+s/2}(\partial M) \to H^{t+s/2}(\partial M)$ belongs to $PS(0,1,0)$. Here the domain and range spaces, even though given the same name, shall be considered Sobolev spaces of sections of different vector bundles, generally speaking of different dimensions. For such a problem, g must satisfy an extra condition, by virtue of the following result.

Exercise 4: If $u \in H_{<1,t>}$ and $Hu = 0$, prove that $(1-R)\, u|_{\partial M} \in H^{s/2+t+1}(\partial M)$. Use the methods of section 1, particularly inequality (6) of that section.

Since $(1-R)\, u$ must enjoy extra smoothness on ∂M, we will crudely suppose that $(1-R)\, u = 0$ on ∂M. Then the problem (3) would be solved if we could find h such that $BRh = g$, since then we need only solve $Hu = 0$, $Ru|_{\partial M} = h$, which can be done if h (and hence g) satisfies a finite number of linear conditions. We take this as motivation for the following theorem. Define $BR: H^{\tau-\delta}(\partial M) \to H^{\tau}(\partial M)$ to be the closed operator with $\mathcal{D}(BR) = \{u \in H^{\tau-\delta} : BRu \in H^{\tau}\}$. Suppose $0 \leq \delta < 1$.

Theorem 3: Suppse the pair $\{H,B\}$ satisfies the following condition

(**) $\qquad BR: H^{\tau-\delta}(\partial M) \to H^{\tau}(\partial M)$ has closed range, of finite codimension. Then for any $g \in H^{t+s/2}(\partial M)$ satisfying a certain finite number of linear conditions, there is a solution $u \in H_{<1,t-\delta>}$ of (3).

Proof: If $E = \{u \in H_{<1,t-\delta>} : Hu = 0\}$, we are to show that closed linear maps $\beta: E \to H^{t+s/2}(\partial M)$ given by $u \to Bu|_{\partial M}$ has closed range of finite codimension.

Let the range of BR be F, a closed finite codimensional subspace of $H^{s/2+t}(\partial M)$. There exists a continuous map $T: F \to H^{\tau-\delta}$ such that $BRTg = g$, $g \in F$. Now if $g \in F$ and if we can solve the equation

$$Hv = 0$$

$$Rv|_{\partial M} = RTg \in H^{s/2+t-\delta}$$

for $v \in H_{<1,t-\delta>}$, which we can if RTg satisfies a finite number of linear conditions, by theorem 2, then $v \in E$, and

$$Bv|_{\partial M} = BRv + B(1-R)v$$

$$= BRTg + B(1-R)v$$

$$= g + B(1-R)v$$

Thus $\beta v - g \in H^{s/2+t-\delta+1}$, so by proposition 2, β has closed range of finite codimension.

Exercise 5: Suppose condition (**) is satisifed. Show that one can solve

$$Hu = f$$

$$Bu = g$$

for $u \in H_{<1,t-\delta>}$, given $f \in H_{(0,t)}$, $g \in H^{t+s/2}(\partial M)$, provided $\{f,g\}$ satisifies a finite number of linear conditions.

If both condition (*) of section 1 and the above condition (**) are satisfied, $u \to \{Hu,Bu\}$ leads to a closed, densely defined Fredholm map of $H_{<1,t-\delta>}$ into $H_{(0,t)} \oplus H^{t+s/2}(\partial M)$. Actually, this coincidence only seems to occur for $\delta = 0$, the most important case, which we discuss in the next section.

We remark that there is a pseudo differential operator, differing from R by a member of $PS(-1,1,0)$, which is a projection onto $\{g \in H^{t+s/2}(\partial M):\ g = u|_{\partial M},\ Hu = 0\}$, the set of "Cauchy data" of solutions to the homogeneous equation $Hu = 0$. For this result, see [33], [66], and [73]. In the results here, we have not made use of this fact.

Exercise 6: Translate theorem 3 and exercise 5 into an existence theorem (modulo finitely many linear conditions) for boundary value problems for elliptic operators on general manifolds with boundary. Prove such a theorem. (You might find proposition 2 useful.)

3. Regular boundary value problems.

Let L be an elliptic operator on M. We may suppose that L maps sections $u \in \Gamma(M,E_0)$ of a vector bundle E_0 into sections $Lu \in \Gamma(M,F)$ of a vector bundle F, of the same fiber dimension as E_0. The process described in section 1 leads to a pseudo differential operators $R:\Gamma(\partial M, E) \to \Gamma(\partial M,E)$, and boundary operators B_j lead to an operator $B:\Gamma(\partial M,E) \longrightarrow \Gamma(\partial M,G)$, where G is another vector bundle, $E = E_0 \otimes \mathbb{C}^m$. Consider the following two conditions:

(a) $\quad |g|_\tau^2 \le C|(1-R)g|_\tau^2 + C|B_g|_\tau^2 + C|g|_{\tau-1}^2 \qquad g \in C^\infty(\partial M)$

(b) $\quad BR: H^\tau(\partial M,E) \to H^\tau(\partial M,G)$ has closed range, of finite codimension.

These are conditions (*) and (**) of section 1 and 2, respectively, with $\delta = 0$. As we have seen, (a) leads to a regularity theorem, and (b) leads to an existence theorem, modulo finitely many linear conditions.

Def: If (a) and (b) are both satisfied, the boundary value problem

$$Lu = f$$

$$B_j u = g_j \qquad 1 \le j \le \nu$$

is called a regular elliptic boundary value problem.

The purpose of this section is to give more explicit conditions for regularity. Let b be the principal symbol of B, r that of R, homogeneous of degree zero in ξ.

Proposition 1: Consider the following conditions.

(1) For each $(x_0, \xi_0) \in T^*(\partial M) \setminus 0$, there is no $v \in E_{x_0}$ such that

$$\begin{cases} v - r(x_0, \xi_0)\, v = 0 \\ \quad b(x_0, \xi_0)\, v = 0 \end{cases}$$

(2) $b(x_0,\xi_0)\, r(x_0,\xi_0) : E_{x_0} \to G_{x_0}$ is surjective, for each $(x_0,\xi_0) \in T^*(\partial M) \setminus 0$.

Then (1) implies (a) and (2) implies (b).

Proof: Condition (1) says that $(1-R^*)(1-R) + B^*B$ is strongly elliptic, from which (a) follows easily. Condition (2) says that $BR(BR)^*$ is elliptic, as an operator on $H^{\tau}(\partial M, G)$, hence Fredholm, from which (b) follows easily.

We want to make these conditions even more explicit by relating them directly to the symbols of L and B_j.

Proposition 2: For given $(x_0,\xi_0) \in T^*(\partial M) \setminus 0$, the following are equivalent.

(1) There is no $v \in E$ such that $v - r(x_0,\xi_0)\, v = 0$ and $b(x_0,\xi_0)\, v = 0$.

(2) There is no solution of the ODE

$$\frac{d}{dt}\phi + \sigma_{K_0(0)}(x_0,\xi_0)\,|\xi_0|\,\phi = 0$$

$$b(x_0,\xi_0)\,\phi(0) = 0$$

with ϕ bounded as $t \to +\infty$, except $\phi \equiv 0$.

(3) There is no solution to the ODE

$$\frac{d^m}{dt^m} \Phi + \sum_{j=0}^{m-1} A_j(x_0,0,\xi_0) \frac{d^j}{dt_j} \Phi = 0$$

$$\tilde{B}_j (x_0,\xi_0, \frac{d}{dt}) \Phi (0) = 0$$

with Φ bounded on $t \to +\infty$, except $\Phi \equiv 0$. Here $\tilde{B}_j(x_0,\xi,\tau)$ is the principal symbol of B_j, and $\frac{d}{dt}$ is substituted for τ in this polynomial.

Propositon 3: For given $(x_0,\xi_0) \in T^*(\partial M)\setminus 0$, the following are equivalent.

(1) $b(x_0,\xi_0)\ r(x_0,\xi_0) : E_{x_0} \to G_{x_0}$ is onto.

(2) There exists a solution to the ODE

$$\frac{d}{dt} \phi + \sigma_{K(0)} (x_0,\xi_0) \ |\xi_0|\ \phi = 0$$

$$b(x_0,\xi_0)\ \phi (0) = \eta$$

for any $\eta \in G_{x_0}$, which is bounded as $t \to +\infty$.

(3) There exists a solution to the ODE

$$\frac{d^m}{dt_m} \Phi + \sum_{j=0}^{m-1} A_j(x_0,0,\xi_0) \frac{d^j}{dt^j} \Phi = 0$$

$$\tilde{B}_j(x_0,\xi_0, \frac{d}{dt}) \Phi (0) = \eta_j$$

for any $\eta_j \in E_{j,x_0}$, where $B_j : \Gamma(M,E) \to \Gamma(M,E_j)$, such that Φ is bounded as $t \to +\infty$.

The proofs of propositions 2 and 3 are practically identical, and very easy. In each case, (1) is equivalent to (2) by virtue of the exponential representation of all the solutions to

$$\frac{d}{dt}\phi + \sigma_{K(0)}(x_0,\xi_0)\ |\xi_0|\ \phi = 0$$

if we recall that $r(x_0,\xi_0)$ is defined to be the projection onto the sum of the generalized eigenspaces of $\sigma_{K(0)}(x_0,\xi_0)$ corresponding to eigenvalues with positive real part, annihilating the other eigenspaces. Finally, (2) and (3) are equivalent because of the equivalence between the nth order

ODE $\dfrac{d^m}{dt^m}\ \Phi + \displaystyle\sum_{j=0}^{m-1} A_j(x_0,0,\xi_0)\ \frac{dj}{dtj}\ \Phi = 0$ and the first order system to which it is reduced.

If the operators $A_j(x,t,D_x)$ are all scalars, we will give another, more algebraic, characterization of regularity. If $L_m(x_0,0,\xi,\tau)$ is the principal symbol of L then none of the roots $\tau_1,\ldots,\tau_m$ of $L_m(x_0,0,\xi,\tau) = 0$ are real, if $\xi \neq 0$. Let

$$M^+(x_0,\xi_0,\tau) = \prod_{k=1}^{j} (\tau - \tau_k(x_0,\xi_0))$$

where we assume $\tau_1(x_0,\xi_0),\ldots,\tau_j(x_0,\xi_0)$ are all the roots with positive imaginary part. Let $\tilde{B}_j(x,t,\xi,\tau)$ be the principal symbol of B_j.

If $\mathbb{C}[\tau]$ denotes the ring of polynomials in τ, with complex coefficients and if $(M^+(x_0,\xi_0,\tau))$ denotes the ideal generated by $M^+(x_0,\xi_0,\tau)$, then $\mathbb{C}[\tau]/(M^+(x_0,\xi_0,\tau))$ is a finite dimensional vector space over $\mathbb{C}$, for each $(x_0,\xi_0) \in T^*(\partial M)\backslash 0$.

Proposition 4: All the conditions of proposition 2, if L and B_j are scalar operators, are equivalent to

(a') $\{\tilde{B}_j(x_0,\xi_0,\tau) : 1 \leq j \leq v\}$ spans $\mathbb{C}[\tau]/(M^+(x_0,\xi_0,\tau))$,

and all the conditions of proposition 3 are equivalent to

(b') $\{\tilde{B}_j(x_0,\xi_0,\tau) = 1 \leq j \leq v\}$ is a linearly independent set in

$\mathbb{C}[\tau]/(M^+(x_0,\xi_0,\tau))$.

Proof: Φ satisfies $\dfrac{d^m}{dt^m} \Phi + \displaystyle\sum_{j=1}^{m-1} A_j(x_0,0,\xi_0) \dfrac{d^j}{dt^j} \Phi = 0$

and is bounded as $t \to +\infty$, if and only if $M^+(x_0,\xi_0, \dfrac{d}{dt})\Phi = 0$. The rest of the argument is left to the reader.

In practice, it is usually as easy to use properties (3) of proposition 2 and 3 to verify regularity as to check properties (a') and (b') above. There are, however, two interesting results which proposition 4 yields immediately.

First of all, suppose L and B_j are scalar differential operators leading to a regular boundary value problem. Since $\tilde{B}_j(x_0,-\xi_0,-\tau) = \pm \tilde{B}_j(x_0,\xi_0,\tau)$, it follows that $\{\tilde{B}_j(x_0,\xi_0,\tau) : 1 \leq j \leq v\}$ is a basis of $\mathbb{C}[\tau]/(M^+(x_0,-\xi_0,-\tau))$ as well as of

$\mathbb{C}[\tau]/(M^+(x_0,\xi_0,\tau))$. In particular it follows that $M^+(x_0,\xi_0,\tau)$ and $M^+(x_0,-\xi_0,\tau)$ have the same degree, for each $\xi_0 \neq 0$. More generally, we have the following.

Def: L is properly elliptic if L is scalar and the degree of $M^+(x_0,\xi_0,\tau)$ is independent of $(x_0,\xi_0) \in T^*(\partial M)\setminus 0$.

Since the roots with positive imaginary part of $L_m(x_0,0,\xi,\tau) = 0$ are also the roots with negative imaginary part of $L_m(x_0,0,-\xi,\tau) = 0$, it follows that a properly elliptic operator L has even order $m = 2\mu$, and $M^+(x_0,\xi_0\tau)$ has order μ, for each $(x_0,\xi_0) \in T^*(\partial M)\setminus 0$. If $\dim M \geq 3$, the assumption that L is properly elliptic is no restriction.

Exercise 7: Show that any elliptic operator L with complex coefficients on a manifold M of dimension $n \geq 3$ is properly elliptic.

Hint: The number of roots τ with positive imaginary part of $L_m(x,t,\xi,\tau)$ is locally constant, for $(x,t) \in M$, $\xi \in \mathbb{R}^{n-1}\setminus\{0\}$. But this set is connected, if $n \geq 3$.

Exercise 8: If $\dim M = 2$ and there is a boundary value problem for L satisfying the conditions of proposition 4, show that L must be properly elliptic.

Proposition 5: If L is a properly elliptic operator of order 2μ, then the Dirichlet problem $Lu = f$, $u|_{\partial M} = g_0$, $\frac{\partial}{\partial\nu} u|_{\partial M} = g_1, \ldots, \frac{\partial^{\mu-1}}{\partial\nu^{\mu-1}} u|_{\partial M} = g_{\mu-1}$ is a regular boundary value problem.

Proof: Here, ν can be any vector field transversal to ∂M. In proper coordinates, $\nu = \frac{\partial}{\partial t}$. The proof amounts to observing that $\{1, \tau, \ldots, \tau^{\mu-1}\}$ must be a basis of $\mathbb{C}[\tau]/(M^+(x_0, \xi_0, \tau))$ under these circumstances.

Exercise 9: Consider $L = (\frac{\partial}{\partial \bar{z}})^2 = (\frac{\partial}{\partial x} + i \frac{\partial}{\partial y})^2$, a second order elliptic operator on $\mathbb{R}^2$. Show that L is not properly elliptic. Verify that the Dirichlet problem on the disc for L is not regular by constructing an infinite dimensional space of solutions to $Lu = 0$, $u|_{S^1} = 0$.

Exercise 10: Let E and B be functions on $\Omega \subset \mathbb{R}^3$ taking values in $\mathbb{R}^3$. Show that the following two elliptic boundary value problems are regular; ν is the unit normal to $\partial\Omega$.

(i)
$$\Delta E = F$$
$$E \times \nu = 0 \text{ on } \partial\Omega$$
$$\operatorname{div} E = 0 \text{ on } \partial\Omega$$

(ii)
$$\Delta B = F$$
$$B \cdot \nu = 0 \text{ on } \partial\Omega$$
$$\nu \times \operatorname{curl} B = 0 \text{ on } \partial\Omega$$

The reader who has seen some physics might note a connection between these boundary value problems and Maxwell's equations for electromagnetic waves in a region bounded by a perfect conductor.

Exercise 11: Let (L,B_j) and (M,C_j) be regular elliptic boundary value problems. Let $\mathcal{D}(L) = \{u \in H^m(M) : B_j u = 0$ on $\partial M\}$, and let $\mathcal{D}(M)$ be defined similarly. Suppose L and M are formally adjoint in the sense that $(Lu,v) = (u,Mv)$ for all smooth u with $B_j u = 0$ on ∂M, smooth v with $C_j v = 0$ on ∂M. Prove that $L^* = M$. In particular, a formally self adjoint boundary problem leads to a self adjoint operator on $L^2(M)$. (See Lions and Magenes [55].)

Exercise 12: If (L, B_j) defines a regular elliptic boundary value problem, let $x_0 \in \partial M$ and suppose $u \in L^2(M)$ is such that Lu and $B_j u|_{\partial M}$ are smooth in some neighborhood of x_0. Prove that u is smooth in a neighborhood of x_0.

Hint: If ϕ is a smooth function supported by a small neighborhood of x_0, note that $L(\phi u) = \phi Lu + [L,\phi]u$, and $[L,\phi]$ has lower order. Thus a priori estimates leads to $u \in H^1$ in a neighborhood of x_0. Repeat this reasoning. (See Lions and Magenes [55].)

4. A subelliptic estimate; the oblique derivative problem.

Let $q(x,D)$ be a pseudo differential operator of order 1 on a compact manifold Ω. If q is elliptic, we have the inequality $\|u\|_1^2 \le C\|q\,u\|_0^2$. In this section we will look at a condition which will lead to a "sub elliptic" estimate $\|u\|_{1/2}^2 \le C\|q\,u\|_0^2 + C\|u\|_0^2$, involving a loss of 1/2 derivative in the smoothness of u.

Assume that the principal symbol $q(x,\xi)$ of q is scalar, and homogeneous of degree 1 in ξ. Then $C = [q^*,q] = q^*q - qq^*$ has real valued principal symbol $C(x,\xi) = i\sum_{j=1}^{n}\left(\frac{\partial q}{\partial x_j}\frac{\partial \bar{q}}{\partial \xi_j} - \frac{\partial \bar{q}}{\partial x_j}\frac{\partial q}{\partial \xi_j}\right)$, in local coordinates, chosen to preserve the volume form on Ω. (Exercise: prove you can always do this.)

Def: If $q \in PS(1,1,0)$ is as above, we say q is subelliptic if $C(x_0,\xi_0) > 0$ whenever $q(x_0,\xi_0) = 0$.

The following theorem is due to Hörmander. Our proof follows Neri [61].

Theorem: If $q \in PS(1,1,0)$ is subelliptic on Ω, then we have the estimate $\|u\|_{1/2}^2 \le C\|qu\|_0^2 + C\|u\|_0^2$

Proof: The hypothesis guarantees that $\eta\Lambda^{-1}q^*q + [q^*,q]$ is a strongly elliptic operator of order 1, if $\eta > 0$ is large enough. Now

$$\begin{aligned}\|q\,u\|_0^2 &= (qu,qu) = (q^*qu,u)\\ &= (qq^*u,u) + ([q^*,q]u,u)\\ &\geq ([q^*,q]u,u)\\ &= (\{\eta q^*\Lambda^{-1}q + [q^*,q]\}u,u) - \eta(\Lambda^{-1/2}qu,\Lambda^{-1/2}qu)\\ &\geq C\,\|u\|_{1/2}^2 - C\|u\|_0^2 - C\,\|q\,u\|_{-1/2}^2\,,\end{aligned}$$

by Gårdings inequality. By means of the inequality

$$\|q\,u\|_{-1/2}^2 \leq \varepsilon\,\|q\,u\|_0^2 + C(\varepsilon)\,\|q\,u\|_{-1}^2 \leq \varepsilon\|q\,u\|_0^2 + C(\varepsilon)\,\|u\|_0^2\,,$$

the result follows.

The more important subelliptic estimates for systems are also much more difficult. We refer the reader to Hörmander [33].

Exercise 13: Let q be subelliptic of order 1, $u \in \mathcal{D}'(\Omega)$, $qu \in H^{\tau}(\Omega)$. Using the Friedrichs mollifier technique, prove that $u \in H^{\tau+1/2}(\Omega)$

Exercise 14: Let $q \in PS(1,1,0)$ and suppose q^* is subelliptic. Define the closed linear operator $Q:H^{\tau-1/2}(\Omega) \to H^{\tau}(\Omega)$ by $\mathfrak{D}(Q) = \{u \in H^{\tau-1/2} : qu \in H^{\tau}\}$, $Qu = qu$. Prove that Q has closed range of finite codimension.

It turns out that if the inequality $C(x_0,\xi_0) < 0$ holds at a single point where $q(x_0,\xi_0) = 0$, no smoothness result can hold. See [33]. Since this is the case for the adjoint of a subelliptic (non-elliptic) operator, we never get Fredholm operators in this manner.

We apply this above result to the oblique derivative problem $\Delta u = 0$, $Lu = f$ on the boundary, when L is a vector field.

For convenience, we will work on the half space $\mathbb{R}^{n+1}_+ = \{(x,t) \in \mathbb{R}^{n+1} : t \leq 0\}$.

Rather than using the general theory of section 1, we will use the Poisson integral to reduce this to a pseudo differential equation on the boundary, like we did in the first chapter.

If $f \in H^s(\mathbb{R}^n)$, the Poisson integral

$$u(x,t) = PI(f) = (2\pi)^{-n} \int_{\mathbb{R}^n} e^{i<x,\xi>-t|\xi|} \hat{f}(\xi)\, d\xi$$

solves the Dirichlet problem $\Delta u = 0$ in $\mathbb{R}^{n+1}_+$, $u|_{t=0} = f$.

Now if u solves the oblique derivative problem

(1)
$$\Delta u = 0$$
$$Bu = a_0(x) \frac{\partial u}{\partial t} + \sum_{j=1}^{n} a_j(x) \frac{\partial u}{\partial xj} = q \quad \text{on } \mathbb{R}^n{}_1$$

suppose $u = PIf$. Then we have

$$Bu|_{\mathbb{R}^n} = (2\pi)^{-n} \int \{-a_0(x)\ |\xi| + i \sum a_j(x)\ \xi_j\} e^{i<x,\xi>} \hat{f}(\xi)\, d\xi$$
$$= Tf$$

where $T \in PS(1,1,0)$ and $\sigma_T(x,\xi) = a_0(x)|\xi| + i \sum_{j=1}^{n} a_j(x)\, \xi_j$.

If a_0 never vanishes, the T is elliptic of order 1, and (1) is a regular elliptic boundary value problem. If $n = 1$, a_j real, it suffices that the $a_0(x), \ldots, a_n(x)$ have no common zero, for T to be elliptic of order 1. This is the situation we saw in Chapter I. However, if $n \geq 2$ and a_0 vanishes at places, then obviously T cannot be elliptic. But if we have the estimate

$$(2) \qquad |Tg|^2 \geq C|g|^2_{\tau-1/2} - C|g|^2_{\tau-1}$$

then we can derive the a priori inequality

$$\|u\|^2_{s-1/2} \leq C\,\|\Delta u\|_{s-2} + C\,\|Bu\|^2_{s-3/2} + C\,\|u\|^2_{s-1}$$

and the corresponding regularity theorem, as before. As we have seen, (2) will hold if $\sigma_T(x_0,\xi_0) = 0$ implies $\sigma_{[T^*,T]}(x_0,\xi_0) > 0$. On the other hand, a corresponding existence theorem will hold if $\sigma_{[T^*,T]}(x_0,\xi_0) < 0$ at such points.

Since the principal symbol of $[T^*,T]$ is

$$\frac{1}{i}\sum_{j=1}^{n}\left\{\frac{\partial q}{\partial x_j}\frac{\partial \bar{q}}{\partial \xi_j} - \frac{\partial q}{\partial \xi_j}\frac{\partial \bar{q}}{\partial x_j}\right\} \quad \text{if} \quad q = \sigma_T, \quad \text{in the}$$

special case $a_0(x) = x_1$, a_j = real const. for $1 \leq j \leq n$, we see that

$$\sigma_{[T^*,T]}(x,\xi) = 2a_1\,|\xi|$$

Thus in this case we get smoothness if $a_1 > 0$ and existence if $a_1 < 0$.

5. References to further work.

The Dirichlet and Neumann ($\frac{\partial}{\partial \nu} u = g$ on $\partial\Omega$) problems for harmonic functions, which started this theory off, are very old. The reduction of such problems to integral equations was started about the turn of the century, and this led to singular integral operators.

General regular elliptic boundary value problems were studied by Lopatinsky and Agmon, Douglas, and Nirenberg. (See [2].) It was known that the oblique derivative problem wasn't regular, if the direction of differentiation was tangent to the boundary. Another non regular boundary value problem, a system, known as the " $\bar{\partial}$-Neumann problem" arose from the theory of several complex variables. The method of section 4 won't handle this; one needs subelliptic estimates for systems proved in [33] to use our technique. For another treatment, see Morrey [59]. For a general treatment of second order elliptic boundary value problems leading to subelliptic estimates, see Kohn and Nirenberg [45].

With the exception of exercise 11, we have made no mention of the spectral properties of elliptic operators. See Agmon [1]. One interesting classical problem is the distribution of eigenvalues of an elliptic operator. See [1]; also see [39], where the powerful tool of Fourier integral operators is used.

For results on elliptic operators with analytic coefficients, see [59].

Chapter VI. Propagation of singularities; wave front sets

1. The wave front set of a distribution.

We say that $(x_0,\xi_0) \in T^*(M)\setminus 0$ is non characteristic for $A \in PS(m,1,0)$ if and only if there is a conical neighborhood(*) U of (x_0,ξ_0) in $T^*(M)\setminus 0$ on which $|\sigma_A(x,\xi)| \geq C|\xi|^m, |\xi| \geq 1$. Otherwise, (x_0,ξ_0) is characteristic for A, and we write $(x_0,\xi_0) \in \gamma(A)$. Thus $\gamma(A)$ is a closed conic subset of $T^*(M)\setminus 0$.

Def: If $u \in \mathcal{E}'(M)$, the wave front set of u is

$$WF(u) = \bigcap_{\substack{A \in PS(0,1,0)\\ Au \in C^\infty(M)}} \gamma(A) \subset T^*(M)\setminus 0.$$

The wave front set, the intersection of the characteristics of all pseudo differential operators which smooth u out, is hence a closed conic subset of $T^*(M)\setminus 0$. We relate this to the familiar notion of the singular support of u, the smallest closed set outside of which u is smooth. Let $\pi: T^*(M) \to M$ be the usual projection of the vector bundle $T^*(M)$ onto its base space.

Proposition 1: $\pi(WF(u))$ = sing supp u, for all $u \in \mathcal{E}'(M)$.

Proof: if $x_0 \notin$ sing supp u, there is a $\phi \in C_0^\infty(M)$, $\phi = 1$ in a neighborhood of x_0, $\phi u \in C_0^\infty(M)$. Since $(x_0,\xi) \notin \gamma(\phi)$ for any $\xi \neq 0$, it follows that $(x_0,\xi) \notin WF(u)$; hence $\pi(WF(u)) \subset$ sing supp u.

On the other hand, if $x_0 \notin \pi(WF(u))$, then for any $\xi \neq 0$, there is a $q \in PS(0,1,0)$ such that $(x_0,\xi) \notin \gamma(q)$ and $qu \in C^\infty$.

(*) A conic subset $S \subset T^*(M)$ is a set that contains the ray $\{(x_0, r\xi_0) : 0 < r < \infty\}$ through each point $(x_0,\xi_0) \in S$.

Thus we can construct finitely many $q_j \in PS(0,1,0)$ such that $q_j u \in C^\infty$ and each (x_0,ξ), $|\xi| = 1$, is non characteristic for some q_j. Then $\bar{q} = \sum_{j=1}^{\nu} q_j^* q_j$ is elliptic in a neighborhood of x_0, while $\bar{q}u \in C^\infty$. Hence u is smooth in a neighborhood of x_0.

Thus the wave front set of u emerges as an object the knowledge of which tells us about sing supp u, and more. This new object is easier to work with than sing supp u; one might guess this to be the case since the proposition connecting the two has already used the regularity theorem for elliptic operators.

We now define an analogous concept for pseudo differential operators.

Def: Let U be an open conic subset of $T^*(M)\setminus 0$. We say $p \in S^m_{1,0}(M)$ has order $-\infty$ on U if for each closed conic subset K of U with $\pi(K)$ compact we have an estimate, for each integer N

$$|D_x^\beta \; D_\xi^\alpha \; p(x,\xi)| \le C_{\alpha,\beta,N,K} \, (1 + |\xi|)^{-N}$$

for all $(x,\xi) \in K$.

Def: The essential support of a symbol of a pseudo differential operator p is the smallest conic subset of $T^*(M)\setminus 0$ on the complement of which p has order $-\infty$. We denote this set by $ES(p)$.

Proposition 2: $ES(p(x,D)\, q(x,D)) \subset ES(p) \cap ES(q)$.

Proof: $\sigma_{pq}(x,\xi) \sim \sum_{\alpha \geq 0} \frac{1}{\alpha!} p^{(\alpha)}(x,\xi)\, (iD_x^{\alpha})\, q(x,\xi)$,

from which the proposition is immediate.

Corollary: $ES(p_1 \ldots p_k) \subset ES(p_1) \cap \cdots \cap ES(p_k)$.

Proposition 3: $WF(pu) \subset WF(u) \cap ES(p)$; for $p \in PS(m,1,0)$.

Proof: First we show that $WF(pu) \subset ES(p)$. If $(x_0, \xi_0) \notin ES(p)$, then there is a conic neighborhood U of (x_0,ξ_0) such that $ES(p) \cap U = \emptyset$. Now let q be any pseudo differential operator such that $q(x,\xi)$ is homogeneous of degree zero in ξ, for $|\xi| \geq 1$, $q(x_0,\xi_0) \neq 0$, and supp $q \subset U$. It follows that $ES(q) \cap ES(p) = \emptyset$, so $qp \in PS(-\infty)$. Hence $qpu \in C^{\infty}(M)$, which shows that $(x_0,\xi_0) \notin WF(pu)$.

In order to show that $WF(pu) \subset WF(u)$, let $(x_0,\xi_0) \notin WF(u)$. Then there exists $A \in PS(0,1,0)$ such that $Au \in C^{\infty}$, $(x_0,\xi_0) \notin \gamma(A)$. We will be done if we can find operators $B, C \in PS(0,1,0)$ such that $(x_0,\xi_0) \notin \gamma(B)$ and $CA = Bp$ mod $PS(-\infty)$ for $B(pu) = C(Au) \in C^{\infty}$.

Let A_0 be an elliptic operator with $\sigma_A = \sigma_{A_0}$ in a comic neighborhood U of (x_0,ξ_0), so $ES(A-A_0) \cap U = \emptyset$. Let $B \in PS(0,1,0)$ be such that $(x_0,\xi_0) \notin \gamma(B)$ and $ES(B) \subset U$, let E_0 be a two sided parametrix for the elliptic operator A_0, and set $C = BpE_0$. Then using proposition 2 we see that

$$\begin{aligned} CA &= BpE_0A \\ &= BpE_0A_0 + BpE_0(A-A_0) \\ &= Bp \bmod PS(-\infty) \end{aligned}$$

Since $E_0A_0 = I \bmod PS(-\infty)$ and $ES(B) \cap ES(A-A_0) = \emptyset$. This completes the proof.

As a corollary, we have the following sharper form of the regularity theorem for elliptic operators.

Corollary: If $p \in PS(m,1,0)$ is elliptic, then $WF(pu) = WF(u)$.

Proof: We have seen that $WF(pu) \subset WF(u)$. On the other hand if E is a parametrix for p, we see that $WF(u) = WF(Epu) \subset WF(pu)$

In the next section we will examine the wave front set of a solution u of a partial differential equation $pu = f$. In fact, we will deal with a more precise concept, which we now define.

Def: Let $(x_0,\xi_0) \in T^*(M)\setminus 0$. We say $u \in H^s$ at (x_0,ξ_0) if we can write $u = u_1 + u_2$ with

(i) $u_1 \in H^s(M)$

(ii) $(x_0,\xi_0) \notin WF(u_2)$

Proposition 4: If $u \in \mathcal{E}'(M)$, $Au \in L^2(M)$, with $A \in PS(s,1,0)$, $(x_0,\xi_0) \notin \gamma(A)$, then $u \in H^s$ at (x_0,ξ_0).

Proof: As in the proof of proposition 3, construct $B \in PS(-s,1,0)$ with $ES(BA-I) \cap U = \emptyset$, where U is a small conic neighborhood of (x_0,ξ_0). Then the decomposition $u = BAu + (1-BA)u$ does the trick.

To put a firmer handle on the notions above, we have the following.

Proposition 5: Let $u \in \mathcal{D}'(M)$, $(x_0,\xi_0) \in T^*(M)\setminus 0$. Then $u \in H^s$ at (x_0,ξ_0) if and only if there is a neighborhood U of x_0 in M, a conic neighborhood V of ξ_0 in $\mathbb{R}^n$, and a function $\phi \in C_0^\infty(U)$ such that

(i) $\phi = 1$ in a neighborhood of x_0

(ii) $(\phi u)^\wedge(\xi) \leq C|\xi|^s$ for $\xi \in V$, $|\xi| \geq 1$.

Since we won't use this proposition, we refer to Hörmander [39] for the proof. Or the reader can prove it as an exercise.

Exercise 1: Show that $WF(u) = \bigcap \{\gamma(p) = pu \in C^\infty,\ p \in PS(0,\rho,\delta)\}$ if $0 \leq \delta < \rho \leq 1$. Show that $WF(pu) \subset WF(u)$ for $p \in PS(m,\rho,\delta)$ if $0 \leq \delta < \rho \leq 1$.

Exercise 2: Extend the corollary to proposition 3 to hypoelliptic operators with constant strength.

Exercise 3: For the solution of an equation $p(x,D)u = f \in C^\infty$ with $p(x,D)$ hypoelliptic of constant strength, there are other natural notions of the wave front set of u. Describe one such and prove assertions parallel to the proposition of this section. (Hint: consider other compactifications of $\mathbb{R}^n$.)

Exercise 4: Let $\Omega \subset \mathbb{R}^n$ be a region with a smooth boundary $\partial\Omega$. Let $u = \chi_\Omega$. Show that $WF(u) = \{(x,\xi) : x \in \partial\Omega, \xi \text{ normal to } \partial\Omega\}$.

2. Propagation of singularities, the Hamilton flow.

If p is a real valued function on $T^*(M)$, the vector field $H_p = \sum_{j=1}^{n} \left(\frac{\partial p}{\partial x_j} \frac{\partial}{\partial \xi_j} - \frac{\partial p}{\partial \xi_j} \frac{\partial}{\partial x_j} \right)$ on $T^*(M)$ is called the Hamiltonian vector field of p. We will be particularly interested in the case when $p(x,\xi)$ is homogeneous of degree s in ξ, so is the principal symbol of a pseudo differential operator.

Def: If $p(x,\xi)$ is real valued and homogeneous in ξ, the integral curves of H_p through $p^{-1}(0)$ in $T^*(M)\setminus 0$ are called the null bicharacteristic strips of p.

The projection of a bicharacteristic strip onto the base manifold M is called a bicharacteristic curve. Note, by the way, that $H_p p = 0$, so p is constant on integral curves of H_p.

Exercise 5: Show that $-iH_p q$ is a principal symbol of $[p(x,D), q(x,D)]$.

We are now ready for the principal theorem of this chapter.

Let $P = p(x,D) \in PS(m,1,0)$ have homogeneous principal symbol. Let $\gamma : I \to T^*(M)\setminus 0$ be a null bicharacteristic strip for $H_{\mathrm{Re}\, p}$, and assume $\mathrm{Im}\, p \geq 0$ on a neighborhood of $\gamma(I)$; $I = [t_0, t_1]$. Let $Pu = f$.

Theorem 1: Under the above assumptions, if $f \in H^s$ on $\gamma(I)$ and if $u \in H^{s+m-1}$ at $\gamma(t_0)$, then $u \in H^{s+m-1}$ on $\gamma(I)$.

In particular if $WF(f) \cap \gamma(I) = \emptyset$ and $\gamma(t_0) \notin WF(u)$, then $\gamma(I) \cap WF(u) = \emptyset$. If P has real principal symbol, we can reverse the time direction in this propostion and conclude that $WF(u) \setminus WF(f)$ is contained in $\gamma(P)$ and is invariant under the flow of H_p.

Proof of theorem: We can multiply P on the left by an elliptic operator E of order $1-m$, to get $P_0 = EP \in PS(1,1,0)$. Now $\gamma(P_0) = \gamma(P)$, and $H_{P_0} = eH_p$ on $\gamma(P)$, so we can assume that $P \in PS(1,1,0)$. We can also assume that $u \in H^{s-1/2}$ on $\gamma(I)$ to start off with, as the reader will be able to see from the proof. Also, without loss of generality, $u \in \mathcal{E}'(M)$.

Now let $\mathfrak{M}$ be a set of pseudo differential operators C with real valued symbols satisfying the following two properties

$$\text{(i)} \quad \mathfrak{M} \subset \{C \in PS(s-1,1,0) : ES(C) \subset \Gamma\}$$

$$\text{(ii)} \quad \mathfrak{M} \text{ is a bounded subset of } PS(s,1,0).$$

Here Γ is a convenient small conic neighborhood of $\gamma(I)$ in $T^*(M) \setminus 0$. Write $P = A + iB$ with $A = A^*$, $B = B^*$. Then for $C \in \mathfrak{M}$ we have $\operatorname{Im}(Cf,Cu) = \operatorname{Re}(BCu,Cu) + \operatorname{Im}([C,A]u,Cu) + \operatorname{Re}([C,B]u,Cu)$. Our plan of attack is to estimate this from below. Let us consider the three terms on the right separately.

I. Since $\sigma_B \geq 0$ on Γ, we have $\operatorname{Re}(BCu,Cu) \geq -C_1 \|Cu\|_0^2 - C_2$ via the sharp Gårding inequality, with C_1, C_2 independent of $C \in \mathfrak{M}$.

II. If $c(x,\xi)$ is the symbol of C, then $-ic\,H_c b$ is a principal symbol of $C^*[C,B]$; this symbol is pure imaginary. Hence

Re $([C,B]u, Cu) \geq - C_3$ with C_3 independent of $C \in \mathfrak{M}$.

III. A principal symbol of $C^*[C,A]$ is $i\, c\, H_a c = \frac{i}{2} H_a c^2 = \frac{i}{2}\{a,c^2\}$. Hence $\mathrm{Im}([C,A]u, Cu) \geq \frac{1}{2} (\{a,c^2\}(x,D)\, u,u) - C_4$, with C_4 independent of $C \in \mathfrak{M}$.

Adding these up, we get

$$I+II+III \leq \mathrm{Im}(Cf,Cu) \leq \varepsilon \|Cu\|_0^2 + \frac{1}{4\varepsilon} \|Cf\|_0^2 \text{ ; hence}$$

(*) $\mathrm{Re}(e(x,D)u,u) \leq \|Cf\|_0^2 + C_5$, C_5 independent of $C \in \mathfrak{M}$,
where $e(x,\xi) = \{a,c^2\}(x,\xi) - (2C_1 + 1)\, c^2(x,\xi)$.

The strategy we want to use in the following. By constructing C in a clever manner and using (*), we want to find an $r \in PS(s,1,0)$ with

$$\mathrm{Re}(r(x,D)^2\, u,u) < \infty$$

where $r(x,\xi) \geq c|\xi|^s$ on a conic neighborhood of $\gamma(I)$. This would immediately imply our desired conclusion: $u \in H^s$ on $\gamma(I)$.

Our family $\mathfrak{M}$ will consist of operators with symbols

$$c_{\varepsilon,\lambda}\,(x,\xi) = c\, e^{\lambda a_0}\, (1 + \varepsilon a_1^2)^{-1/2}$$

where λ is a conveniently fixed real number and $0 < \varepsilon \leq 1$. The requirements we make are the following:

(a) $c(x,\xi)$ is homogeneous of degree s in ξ; supp $c \subset \Gamma$; $c(x,\xi) \geq 0$.

(b) $H_a c \geq 0$ on $\Gamma \setminus U$, where U is a small conic neighborhood of $\gamma(t_1)$, with strict inequality on $\gamma(I) \setminus U$. ($A(x,\xi) = \sigma_A = \mathrm{Re}\ p$.)

(c) $a_0 \in S^0_{1,0}$; $H_a a_0 = 1$, in a nbd of $\gamma(I)$.

(d) $a_1 \in S^1_{1,0}$; $H_a a_1 = 0$; a_1 not zero on $\gamma(I)$.

Let us show how this may be arranged. We assume that, at $\gamma(t_0)$, H_a is not radial, since otherwise the proposition is trivial. Now we can construct a conic hypersurface Ω in $T^*(M) \setminus 0$ through $\gamma(t_0)$, transversal to H_a in a neighborhood of $\gamma(t_0)$. Let g be an arbitrary positive smooth function supported on a small conic neighborhood of $\gamma(I)$, with $g(x,\xi)$ homogeneous of degree s in ξ, $|\xi| \geq \varepsilon$. Now solve the ordinary differential equation

(1) $$H_a c = g$$

$$c = 0 \text{ on } \Omega$$

Then multiplying by a smooth function identically 1 in a neighborhood of $\gamma(I) \setminus U$ cut c off in a neighborhood of $\gamma(t_1)$, so that (a) and (b) above hold; $c(x,\xi)$ must be homogeneous of degree

s in ξ by the homogeneity of a and g. Similarly we have a_0, a_1 satisfying conditions (c) and (d) above. Hence we have $e_{\lambda,\varepsilon}(x,\xi) = \{a, c^2_{\lambda,\varepsilon}\} - (2C_1-1)\, c^2_{\lambda,\varepsilon} = (\{a,c^2\} + (2-2C_1-1)c^2)$ $e^{\lambda a_0}(1+\varepsilon^2 a_1{}^2)^{-1/2}$. Now fix $\lambda > C_1 + 1$. Hence $(\{a,c^2\} + (2-2C_1-1)c^2)\, e^{\lambda a_0}$ is homogeneous of degree $2s$, ≥ 0 on $\Gamma\setminus U$, and > 0 on $\gamma(I)\setminus U$. Thus we can find r, q homogeneous of degree s, positive, with supp $q \subset U$, and $r^2 \leq (\{a, c^2\} + (2-2C_1-1)c^2)\, e^{\lambda a_0} + q^2$.
If we apply the sharp Gårding inequality to the difference of the two sides of this inequality, multiplied by $(1+\varepsilon^2 a_1{}^2)^{-1/2}$, we get, using (*), $\|r_\varepsilon u\|_0^2 = (r_\varepsilon^2 u, u)$

$$\leq \mathrm{Re}(e_{\lambda,\varepsilon}(x,D)\, u,u) + \|q_\varepsilon u\|_0^2 + C_6$$

$$\leq \|C_{\lambda,\varepsilon} f\|_0^2 + \|q_\varepsilon u\|_0^2 + C_7$$

where C_6, C_7 are independent of ε, $0 < \varepsilon \leq 1$, assuming $u \in H^{s-1/2}$ on $\gamma(I)$. Letting $\varepsilon \to 0$, we get $\|r(x,D)u\|_0^2 \leq M < \infty$, so $u \in H^s$ on $\gamma(I)$, as desired. The proof is complete.

Two remarks on the proof we have just been through: Note that we need only assume the principal part of A is scalar; this observation will be used when we discuss systems. The ordinary differential equation (1) which we had to solve for c is a tool borrowed from geometrical optics, known as the transport equation.

Exercise 6: Let Φ be the fundamental solution of the hyperbolic operator L considered in exercise 5 of Chapter IV (section 4). Prove that $WF(\Phi)$ is contained in the null bicharacteristic strips over the upper half space which pass over the origin.

Hint: Every bicharacteristic strip passes into the lower half space, where $\Phi = 0$; some miss the origin, where $L\Phi$ is singular; some don't.

Exercise 7: Fix $\varphi \in C_c^\infty$, $\varphi(0)=1$. Let ξ_0 be fixed, and

$$p(\xi)= \sum_{n=1}^{\infty} \varphi\left(\frac{1}{n}(\xi - n^2 \xi_0)\right).$$

(a) Show that $p \in S^0_{\frac{1}{2},0}(\mathbb{R}^n)$.

(b) If $v(\xi)= e^{i x_0 \cdot \xi}\, p(\xi)$ and $u=\hat{v}$, snow that

$$WF(u)= \{(x_0, r\xi_0): 0<r<\infty\}.$$

3. Local existence.

Proposition 1: Let $P \in PS(m,1,0)$ have real principal symbol, homogeneous of degree m. If no null bicharacteristic strip lies over a compact set K of M, then $u \in \mathcal{E}'(M)$, $Pu \in C^\infty(M) \Rightarrow u \in C^\infty(M)$.

Proof: By the theorem of the previous section, $WF(u) = \emptyset$.

Proposition 2: Suppose the following three conditions hold;

(a) $P \in PS(m,1,0)$, properly supported

(b) K is a compact subset of M

(c) $u \in \mathcal{E}'(K)$, $Pu \in C^\infty \Rightarrow u \in C_0^\infty(K)$.

If k and s are arbitrary, then there exists an ℓ such that

(1) $\quad \|u\|_k \leq C\|Pu\|_\ell + C\|u\|_s \quad u \in C_0^\infty(K)$

Proof: $C = \{u \in H_0^s(K) = Pu \in C^\infty(M)\}$ is a Frechet space, with the obvious topology. The inclusion $C_0^\infty(K) \overset{j}{\rightarrow} E$ is continuous and one-one; by hypothesis (c) it is onto. The conclusions hence follows from the open mapping theorem.

For the local existence theorem we derive from this, we will assume that $p(x,D)$ is a differential operator of order m, with real principal part. Define a closed, densely defined operator $P:H_0^k(K) \to H_0^\ell(K)$ by $\mathcal{D}(P) = \{u \in H_0^k(K):p(x,D)u \in H_0^\ell(K)\}$, $Pu = p(x,D)u$. If inequality (1) holds, P has closed range and finite dimensional kernel. Hence P^* has closed range of finite codimension. Since $\mathcal{D}(P^*) \subset H^{-\ell}(K)$, which is not a space of distribution on M (but rather a space of équivalence classes of such distributions) some care must be taken in interpreting this last result. However, any $u \in H^{-\ell}(K)$ defines a distribution in $\mathring{K}$ the interior of K, and $P^*u = p(x,D)^*u$ on $\mathring{K}$.

If $p(x,D)$ has real principal part, $p(x,D)^*$ has the same principal part, hence the same set of characteristic and null bicharacteristic strips. Thus the roles of $p(x,D)$ and $p(x,D)^*$ can be reversed in the above argument. We now sum up what we have so far.

Proposition 3: Let $p(x,D)$ be a differential operator with real principal part, K a compact subset of M, and suppose no null bicharacteristic strip lies entirely over K. Then there is a finite dimensional linear space $E \subset C_0^\infty(K)$ with the property that for any $f \in \mathcal{E}'(K)$, $f \in H^{-k}$, $f \perp E$, we can find $u \in H^{-\ell}(M)$ satisfying the equation

$$Pu = f$$

on $\mathring{K}$.

We are now ready to state and prove our local existence theorem.

Theorem 1: Let $p(x,D)$ be a differential operator of order m, with real principal part, and assume that no null bicharacteristic strip (of the principal part) lies over a single point. Then for any $f \in \mathcal{D}'(M)$, $p_0 \in M$, there exists a neighborhood U of p_0 and $u \in \mathcal{D}'(U)$ such that

$$p(x,D)u = f \quad \text{on } U.$$

Proof: We show that there is a compact neighborhood K of P_0 over which no null bicharacteristic strip completely lies. In fact, if such strips γ_ν lie over $K = \{x \in M: |x-P_0| \leq \frac{1}{\nu}\}$, since by homogeneity of H_{p_m} we can assume γ_ν passes through the cosphere bundle of M at some point over K_ν, we have a limiting null bicharacteristic strip γ lying over P_0, a contradiction.

Thus, for each ν sufficiently large there is a finite dimensional linear space $E_\nu \subset C_0^\infty(K_\nu)$ such that $p(x,D)u = f$ can be solved, say on $K_{\nu+1}$, for all $f \in \mathcal{D}'(M)$ with $f \perp E_\nu$. It remains only to show that $E_\nu = 0$ for sufficiently large ν.

But clearly $E_j \supset E_{j+1} \supset \ldots$, so the dimensions, being finite and decreasing, must stabilize: $E_k = E_{k+1} = \ldots$ for a certain k. But since $f \in E_k \Longrightarrow \text{supp } f \subset K_\nu$ for all $\nu \geq k$, we have $f = 0$ since $f \in C^\infty$. Thus $E_k = 0$, and the proof is complete.

Note in particular that we can take U independent of f.

The argument we have given, via proposition 1 and 2, has used only the fact that $WF(u) \setminus WF(p(x,D)u)$ is invarient under the flow generated by H_{p_m}. Actually, the theorem of the last section yields the following strong form of inequality(1), when $p(x,D)$ satisfies the hypotheses of proposition 1:

(1') $$\|u\|_{\ell+m-1} \leq C\,\|Pu\|_\ell + C\,\|u\|_s \qquad u \in C_0^\infty(K)$$

Exercise 8: Prove that if $p(x,D)$ satisfies the conditions of theorem 1, then each $P_0 \in M$ has a neighborhood U on which the equation

$$p(x,D)u = f$$

can be solved for $u \in H^{m-1+s}$, given $f \in H^s(M)$.

Exercise 9: In the context of exercise 5, can you prove that $p(x,D)u = f$ is locally solvable for $u \in C^\infty$, given $f \in C^\infty(M)$?

Exercise 10: Show that the operator $\frac{\partial}{\partial\theta}$ on the annular $\Omega = \{z \in \mathbb{C}: 1 \leq |z| \leq 2$ has null bicharacteristic strips lying over compact sets. Check $\frac{\partial}{\partial\theta}$ against proposition 3.

Such a local solvability result as we just derived was first proved by Hörmander; see [31]. In [33] he introduced the sharp Gårding inequality for such a purpose. Our treatment here follows [35].

For differential operators with complex coefficients in the principal part, results could be derived from the theorem of the last section, but they are less complete than they are in the real case. Serious obstructions to local solvability in this case were first noticed by Lewy. See Chapter 6 of [31]. Nirenberg and Treves [62] have given necessary and sufficient conditions for local solvability in the case of analytic coefficients. Very recently, Fefferman and Beals have extended the results to the case of merely C^∞ coefficients. See [86].

4. Systems; an exponential decay result.

Theorem 1: Let $p(x,D)$ be a $k \times k$ system in $PS(m,1,0)$. Assume that $q(x,\xi) = \det p(x,\xi)$ has real principal part q_M, of order $M = km$. Let $p(x,D)u = f$, and let $\gamma : I \to T^*(M)\setminus 0$ be a null bicharacteristic strip for q_M; $I = [t_0,t_1]$. If $f \in H^s$ on $\gamma(I)$ and if $u \in H^{s+m-1}$ at $\gamma(t_0)$, then $u \in H^{s+m-1}$ on $\gamma(I)$.

Proof: If ${}^{co}p(x,\xi)$ is the cofactor matrix of $p(x,\xi)$, then ${}^{co}p(x,D)\, p(x,D) = q_M(x,D) + r(x,D)$ where r has order $M-1$. Then $(q_M(x,D) + r(x,D))u = {}^{co}p(x,D)f \in H^{s-m(k-1)}$ on $\gamma(I)$. Since the principal part of $q_M + r$ is scalar Theorem 1 of section 2 yields the desired conclusion.

Suppose now that $L = \frac{\partial}{\partial t} + G(x,D_x)$ is a symmetric first order $k \times k$ hyperbolic system on $M \times \mathbb{R}$, where M is a compact manifold. Suppose in fact that

$$G(x,D_x) + G(x,D_x)^* = -F(x)$$

where F is a positive self adjoint matrix function. We call L a dissipative symmetric hyperbolic operator. If u satisfies $Lu = 0$ and $u(0) \in L^2(M)$, then $u(t) \in L^2(M)$ for all t, and $\frac{d}{dt}\, \|u(t)\|^2 = 2\,\mathrm{Re}(G(x,D_x)u,u) = -2 \int_M F(x)u \cdot u\, dx \leq 0.$

What we aim to show is that, under certain conditions, $u(t)$ decays to zero at an exponential rate: $\|u(t)\| \leq C\, e^{-\alpha t}$; as far as we know, this result is new. For further details, see [92].

Suppose $-F(x) \leq -\eta < 0$ on an open subset U of M. We shall make the following assumption, with $q(x,\xi) = \det G(x,\xi)$:

(A) There is a number T such that every null bicharacteristic strip of q_M in $T^*(M \times \mathbb{R}) \setminus 0$ passes through $T^*(U \times (0,T)) \setminus 0$.

Assumption (A) together with theorem 1 and the closed graph theorem, gives us the following inequality, with $\omega = [0,T] \times U$, $\Omega = [0,T] \times M$: $\|u\|^2_{L^2(\Omega)} \leq C \|Lu\|^2_{L^2(\Omega)} + C\|u\|^2_{L^2(\omega)} + C\|u\|^2_{H^{-s}(\Omega)}$.

Thus if $T: L^2(\Omega) \to L^2(\Omega) \oplus L^2(\omega)$ is the closed linear operator given by $Tu = \{Lu, u|_\omega\}$, $\mathcal{D}(T) = \{u \in L^2(\Omega) : Lu \in L^2(\Omega)\}$, it follows that T has closed range. We make a further assumption:

(B) If T is taken sufficiently large, then $Lu = 0$ on Ω, $u = 0$ on ω implies $u \equiv 0$ on Ω.

Assumption (B) is satisfied, for example, if M is an analytic manifold and L has analytic coefficients, by Holmgren's uniqueness theorem. A result of [69] together with a dimension argument like that used in the proof of the local solvability theorem of the last section can be used to show that assumption (B) holds if G is elliptic and the elliptic operator $\mathcal{G}$ has the unique continuation property.

Granted assumption (B), $T^{-1}: R(T) \to L^2(\Omega)$ is well defined and, by the closed graph theorem, continuous. This yields the inequality

$$\|u\|^2_{L^2(\Omega)} \le C\,\|Lu\|^2_{L^2(\Omega)} + C\,\|u\|^2_{L^2(\omega)} .$$

Now if $Lu = 0$, $u(0) \in L^2(M)$, it follows that

$$\frac{d}{dt}\|u(t)\|^2_{L^2(M)} = -2\int_M FU \cdot u\,dx \le -2\eta\,\|u(t)\|^2_{L^2(U)}$$

$$\therefore\ \|u(T)\|^2_{L^2(M)} \le \|u(0)\|^2_{L^2(M)} - 2\eta\,\|u\|^2_{L^2(\omega)}$$

$$\le \|u(0)\|^2_{L^2(M)} - C\,\|u\|^2_{L^2(\Omega)}$$

$$\le \|u(0)\|^2_{L^2(M)} - CT^2\|u(T)\|^2_{L^2(M)}$$

$$\|u(T)\|^2 \le (1 + CT)^{-1}\,\|u(0)\|^2.$$

From this inequality, exponential decay is a very simple consequence.

Exercise 11: Suppose there is a function $\rho(t) \to 0$ as $t \to +\infty$ such that for $u(0) \in L^2(M)$, $Lu = 0$, $\|u(t)\|_{L^2(M)} \le C\rho(t)$, where C may depend on u. Prove that solutions to $Lu = 0$ with $u(0) \in L^2$ decay exponentially. Hint: Use the uniform boundeness theorem to prove that there is an $\alpha < 1$, $T < \infty$ such that $\|u(T)\|_{L^2} \le \alpha\,\|u(0)\|_{L^2}$ for all such u. Here $L = \frac{\partial}{\partial t} + G(x,D_x)$ can be any hyperbolic operator for which the Cauchy problem is L^2 well posed in the sense that the conclusion of the existence theorem of Chapter IV, section 2, holds.

We should point out that it is known that, for strictly hyperbolic systems, assumption (A) is necessary for exponential decay. See Ralston, Comm. Pure Appl. Math. XXII (1969), 807-823. Assumption (A) basically implies that L is strictly hyperbolic.

Chapter VII. The Sharp Gårding Inequality

In Chapter II we stated, and in the last chapter we used, a sharpened form of Gårding's inequality: if $p(x,\xi) \geq 0$, $p \in S^m_{1,0}$, then

$$\mathrm{Re}(p(x,D)\, u,\, u) \geq -C\, \|u\|^2_{1/2(m-1)},\quad u \in C_0^\infty(K).$$

This result was first proved by Hörmander [33] in the scalar case, which is really all we've used It was extended to the case where $p(x,\xi)$ is a positive matrix valued function by Lax and Nirenberg [51]. Friedrichs [22], using multiple symbols, simplified their proof. **See also [81]**

We shall follow Friedrichs in modifying $p(x,D)$ to make it positive by means of a multiple symbol. However, our treatment of such a multiple symbol will parallel the development in Chapter II, section 3. We can extend the result to positive symbols $p(x,\xi) \in S^m_{\rho,\delta}$, $0 \leq \delta < \rho \leq 1$, with no extra work. This approach was taken in [90]. A further generalization is given in [87].

1. A multiple symbol.

Def: $b(\xi_2,x,\xi_1) \in S^{m_1,m_2}_{\rho,\delta_1,\delta_2}$ if for K compact we have an estimate

$$\left|D^{\gamma}_{\xi_2} D^{\beta}_{x} D^{\alpha}_{\xi_1} b(\xi_2,x,\xi_1)\right| \leq C_{K,\alpha,\beta,\gamma} (1 + |\xi_2|)^{m_2-\rho|\gamma|+\delta_2|\beta|}$$

$$(1 + |\xi_1|)^{m_1-\rho|\alpha|+\delta_1|\beta|},$$

for $x \in K$, $\xi_1,\xi_2 \in \mathbb{R}^n$.

For convenience we will assume that $b(\xi_2,x,\xi_1) = 0$ for x outside some compact set. Let $\hat{b}(\xi_2,\eta,\xi_1) = \int b(\xi_2,x,\xi_1)\, e^{i<x,\eta>} d\eta$.

Def: $b(D,x,D) = B$ is defined by the formula

$$(Bu)^\wedge(\xi_2) = (2\pi)^{-n} \iint b(\xi_2,y,\xi_1)\, e^{i<y,\xi_2-\xi_1>} \hat{u}(\xi_1)\, d\xi_1 dy$$

$$= (2\pi)^{-n} \int \hat{b}(\xi_2,\xi_2-\eta,\eta)\, \hat{u}(\eta)\, d\eta.$$

Note that $b(D,x,D)u = (2\pi)^{-n} \int a(x,\xi_1)\, e^{i<x,\xi_1>} \hat{u}(\xi_1)\, d\xi_1$,

$$\text{where } a(x,\xi) = \int b(\xi_2,y,\xi)\, e^{i<y-x,\xi_2-\xi>} dy\, d\xi_2$$

$$= \int \hat{b}(\xi_2,\xi_2-\xi,\xi)\, e^{-i<x,\xi_2-\xi>} d\xi_2$$

$$= \int \hat{b}(\xi+\zeta,\zeta,\xi)\, e^{-i<x,\zeta>} d\zeta.$$

This suggests that $b(D,x,D)$ is a pseudo differential operator. We now show that such is the case, and give an asymptotic formula for its symbol.

Theorem: $a(x,\xi) \in S^{m}_{\rho,\delta}$ with $m = m_1 + m_2$, $\delta = \delta_1 + \delta_2$, if $0 \le \delta < \rho \le 1$, and we have the asymptotic relation

$$a(x,\xi) \sim \sum_{\alpha \ge 0} \frac{1}{\alpha!} D_x^{\alpha} (i D_{\xi_2})^{\alpha} b(\xi_2,x,\xi)|_{\xi_2=\xi}.$$

Proof: By Taylor's formula we have

$$\hat{b}(\xi + \zeta,\zeta,\xi) = \sum_{|\alpha|<N} \frac{1}{\alpha!} \zeta^{\alpha} D^{\alpha}_{\xi_2} \hat{b}(\xi_2,\zeta,\xi)|_{\xi_2=\xi} + R_N(\zeta,\xi)$$

with $|R_N(\zeta,\xi)| \le C_N \sup_{\substack{|\gamma|=N \\ 0 \le t \le 1}} \left|D^{\gamma}_{\xi_2} \hat{b}(\xi + t\zeta,\zeta,\xi)\right| |\zeta|^N.$

Taking inverse Fourier transforms with respect to ζ yields

$$a(x,\xi) = \sum_{|\alpha|<N} \frac{1}{\alpha!} D^{\alpha}_{x} (iD_{\xi_2})^{\alpha} b(\xi_2,x,\xi)|_{\xi_2=\xi} + \tilde{R}_N(x,\xi)$$

where $\tilde{R}_N(x,\xi) = \int R_N(\zeta,\xi)\, e^{-i<x,\zeta>} d\zeta$. The general term in the sum clearly belongs to $S^{m-(\rho-\delta)|\alpha|}_{\rho,\delta}$. To complete the proof, we need only verify the following facts, in view of Theorem 2 of Chapter II, section 3

(i) $|\tilde{R}_N(x,\xi)| \le C_N (1 + |\xi|)^{m+n-(\rho-\delta)N}$

(ii) $|D^{\alpha}_{\xi} D^{\beta}_{x} a(x,\xi)| \le C_{\alpha,\beta} (1 + |\xi|)^{\mu}$, $u = \mu(\,,\beta)$.

Proof of (i): If $b \in S^{m_1,m_2}_{\rho,\delta_1,\delta_2}$ then

$$\left|D^{\gamma}_{\xi_2} \hat{b}(\xi_2,\eta,\xi_1)\right| \le C_{\gamma,\nu} (1 + |\xi_2|)^{m_2-\rho|\gamma|+\delta_2\nu}$$

$$(1 + |\xi_1|)^{m_1+\delta_1\nu} (1 + |\eta|)^{-\nu}.$$

$$\therefore |R_N(\zeta,\xi)| \le C_N \sup_{\substack{|\gamma|=N \\ 0\le t\le 1}} \left| D^{\gamma}_{\xi_2} \hat{b}(\xi + t\zeta,\zeta,\xi)\right| |\zeta|^N$$

$$\le C_{N,\nu} \sup_{0\le t\le 1} (1 + |\xi + t\zeta|)^{m_2-\rho N+\delta_2\nu}(1 + |\xi|)^{m_1+\delta_1\nu}$$

$$(1 + |\zeta|)^{N-\nu}.$$

With $\nu = N$ we obtain a bound of $C(1 + |\xi|)^{m-(\rho-\delta)N}$ for $|\zeta| \le 1/2\ |\xi|$, and if ν is large we get a bound by any power of $(1 + |\zeta|)^{-1}$ for $|\xi| \le 2\ |\zeta|$. From this, inequality (i) follows.

Proof of (ii): $D^{\alpha}_{\xi} D^{\beta}_{x}\ a(x,\xi)$

$$= \sum_{\alpha_1+\alpha_2=\alpha} C_{\alpha_1,\alpha_2} \int (y-x)^{\alpha_1} (\xi_2-\xi)^{\beta} \{D^{\alpha_2}_{\xi}\ b(\xi_2,y,\xi)\}\ e^{i<y-x,\xi_2-\xi>} dy d\xi_2.$$

Thus we need to bound $\int y^{\epsilon}\ \xi_2^{\gamma} \{D^{\alpha_2}_{\xi}\ b(\xi_2,y,\xi)\}\ e^{i<y-x,\xi_2-\xi>} dy d\xi_2$ by a power of $1 + |\xi|$. This expression is equal to

$$\int \xi_2^{\gamma}\ D^{\alpha_2}_{\xi_1}\ \hat{b}_0(\xi_2,\xi_2-\xi,\xi_1)\Big|_{\xi_1=\xi}\ e^{-i<x,\xi_2-\xi>}\ d\xi_2$$

$$= \int (\xi+\zeta)^{\gamma}\ \{D^{\alpha_2}_{\xi_1}\ \hat{b}_0(\zeta-\xi,\zeta,\xi_1)\Big|_{\xi_1=\xi}\}\ e^{-i<x,\zeta>} d\zeta,$$

where we have set $b_0 = y^{\epsilon}\ b$. The integrand in this last expression is bounded in absolute value by

$$C_{\nu}\ |\xi+\zeta|^{|\gamma|}\ \{1+|\zeta+\xi|\}^{m_2+\delta_2\nu}\ \{1+|\xi|\}^{m-|\alpha|\rho+\delta_1\nu}\ \{1+|\zeta|\}^{-\nu}.$$

Thus it is easy to bound $D^{\alpha}_{\xi} D^{\beta}_{x}\ a(x,\xi)$ by a power of $1 + |\xi|$, and the proof of the theorem is complete.

Excercise 1: Define more general classes of multiple symbols and associated pseudo differential operators, such as $b(x,D,x,D,x,D)$, etc.

Exercise 2: If $a(\xi,x) \in S^{0,m}_{\rho,\sigma,\delta}$, show that $a(D,x) \in PS(m,\rho,\delta)$ is the same operator defined by this method as defined in section 3, chapter II.

If $p(x,D)$ $PS(\mu,\rho,\delta)$, show that $a(D,x)p(x,D) = b(D,x,D)$ where $b(\xi_2,x,\xi_1) = a(\xi_2,x)\ p(x,\xi_1)$.

2. Friedrich's symmetrization.

Suppose $p(x,\xi) \in S^m_{\rho,\delta}$. Let $q(\xi) \geq 0$ be an even function, $q \in C_o^\infty(|\xi| \leq 1)$, and suppose $\int q(\xi)^2\, d\xi = 1$. We define the multiple symbol

$$(1) \qquad b(\xi_2,x,\xi_1) = \int F(\xi_2,\zeta)\, p(x,\zeta)\, F(\xi_1,\zeta) d\zeta$$

where $F(\xi,\zeta) = q((\zeta-\xi)(1+|\xi|^2)^{-\tau/2})(1+|\xi|^2)^{-\tau n/4}$, $\tau = 1/2\ (\rho+\delta)$. Our purpose in doing this is that, if $p(x,\xi) \geq 0$, then $b(D,x,D) = B$ will be a positive pseudo differential operator, close to $p(x,D)$.

In fact, if $p(x,\xi) \geq 0$, then for any decent f,

$$\int f(\xi_1) \cdot b(\xi_2,x,\xi_1)\, \overline{f(\xi_2)}\, d\xi_2 d\xi_1$$

$$= \int \{\int f(\xi_1)\, F(\xi_1,\zeta) d\xi_1\}\, p(x,\zeta)\, \overline{\{\int f(\xi_2)\, F(\xi_2,\zeta) d\xi_2\}}\, d\zeta$$

$$\geq 0 \quad \textbf{since } p(x,\xi) \geq 0.$$

Hence

$$(B\,u,u) = \int \overline{\hat{u}(\xi_2)} \cdot \hat{b}(\xi_2,\xi_1-\xi_2,\xi_1)\, \hat{u}(\xi_1) d\xi_1 d\xi_2$$

$$= \int \hat{u}(\xi_1)\, e^{i<\xi_1,x>} \cdot b(\xi_2,x,\xi_1)\, \overline{\hat{u}(\xi_2)}\, e^{-i<\xi_2,x>} d\xi_1 d\xi_2$$

$$\geq 0 .$$

The next goal is to prove that $b(D,x,D)$ is a pseudo differential operator.

Proposition: $b(\xi_2,x,\xi_1) \in S^{m,o}_{\tau,\delta,o}$.

Proof: An inductive argument shows that $D^\beta_\xi F(\xi,\zeta)$ has the form

$$(2) \qquad D_\xi^\beta F(\xi,\zeta) = (1+|\xi|^2)^{-\tau n/4} \sum_{\substack{|\gamma|\le\beta \\ \gamma_1\le\gamma}} \psi_{\beta,\gamma,\gamma_1}(\xi)\cdot$$

$$\cdot\, ((\zeta-\xi)(1+|\xi|^2)^{-\tau/2})^{\gamma_1}\, D_\sigma^\gamma\, q(\sigma)$$

$$(\text{setting } \sigma = (\zeta-\xi)(1+|\xi|^2)^{-\tau/2})$$

where

$$\psi_{\beta,\gamma,\gamma_1}(\xi) \in S_{1,0}^{-(|\rho| - (1-\tau)|\gamma-\gamma_1|)} \subset S_{\rho,\delta}^{-\tau|\beta|}.$$

Now applying Schwartz' inequality to (1), we get

$$|D_{\xi_2}^\gamma\, D_x^\beta\, D_{\xi_1}^\alpha\, b(\xi_2,x,\xi_1)|$$

$$\le\ (\int|D_{\xi_2}^\gamma\, F(\xi_2,\zeta)D_x^\beta\, p(x,\zeta)|^2 d\zeta)^{1/2}\ (\int|D_{\xi_1}^\alpha\, F(\xi_1,\zeta)|^2 d\zeta)^{1/2}.$$

We can estimate the second factor, using (2) plus the fact that $|(\zeta-\xi)(1+|\xi|^2)^{-\tau/2}| \le 1$ on $\mathrm{supp}\, q((\zeta-\xi)(1+|\xi|^2)^{-\tau/2})$.

$$\int|D_{\xi_1}^\alpha\, F(\xi_1,\zeta)|^2 d\zeta \le C_\alpha(1+|\xi_1|^2)^{-\tau|\alpha|},$$

and similarly we get an estimate for the first term

$$\int|D_{\xi_2}^\gamma\, F(\xi_2,\zeta)D_x^\beta\, p(x,\zeta)|^2 d\zeta$$

$$\le\ C_{\beta,\gamma}(1+|\xi_2|^2)^{-\tau|\gamma|}\int|q_{|\gamma|}(\sigma)D_x^\beta\, p(x,\xi_2+\sigma(1+|\xi_2|^2)^{\tau/2})|^2 d\sigma$$

$$\le\ C'_{\beta,\gamma}(1+|\xi_2|^2)^{m+\delta|\beta|-\tau|\gamma|}$$

with $q_\ell(\sigma) = \max\limits_{|\gamma|\le\ell} |D_\sigma^\gamma\, q(\sigma)|$. The proof is complete.

This proposition shows that $b(D,x,D) \in PS(m,\tau,\delta)$. The crux of the sharp Gårding inequality **is** that, not only is $b(D,x,D) \in PS(m,\rho,\delta)$, but $p(x,D) - b(D,x,D) \in PS(m-(\rho-\delta),\rho,\delta)$. We will prove this in the next section, using the asymptotic formula given in section 1.

A mysterious element in this symmetrization procedure is the necessity for so much care in producing the formula (1). One is tempted to try the **simpler formula.**

$$(3) \qquad b_o(\xi_2,x,\xi_1) = \int q(\xi_2-\zeta)\, p(x,\zeta)\, q(\xi_1-\zeta)\,d\zeta \,.$$

Unfortunately, this does not yield a b_o which belongs to one of our classes of symbols.

Exercise 3: With $q(\xi) = ce^{-|\xi|^2}$ and $p(x,\xi) = 1+|\xi|^2$, evaluate $b_o(\xi_2,x,\xi_1)$ as defined by (3) and show it is not a symbol.

3. The sharp Gårding inequality.

We take b(D,x,D) to be the Friedrich's symmetrization of p(x,D).

Theorem: $b(D,x,D) \in PS(m,\rho,\delta)$, and, more crucially,

$$b(D,x,D) - p(x,D) \in PS(m-(\rho-\delta),\rho,\delta).$$

Proof: We know that $b(D,x,D) = a(x,D)$ is a pseudo differential operator, and we know the asymptotic expansion of $a(x,\xi)$:

$$a(x,\xi) \sim \sum_{\alpha\geq 0} p_\alpha(x,\xi)$$

where $p_\alpha(x,\xi) = 1/\alpha!\, D_x^\alpha (iD_{\xi_2}^\alpha)\, b(\xi_2,x,\xi)|_{\xi_2=\xi}$. The estimates of the previous section show that

$$p_\alpha(x,\xi) \in S^{m-(\rho-\delta)|\alpha|/2}_{\tau,\delta}.$$

The plan here is to show that

$$(*) \qquad p(x,\xi) - \sum_{|\alpha|<N} p_\alpha(x,\xi) = r_N(x,\xi) \in S^{m-(\rho-\delta)}_{\rho,\delta},$$

for N sufficiently large. It is easy to see that this will complete the proof.

Now, using formula (2) of the previous section, write

$$p_\alpha(x,\xi) = \sum_{\substack{|\gamma|\leq|\alpha| \\ \gamma_1\leq\gamma}} \psi_{\alpha,\gamma,\gamma_1}(\xi) \int D_x^\alpha\, p(x,\xi+\sigma(1+|\xi|^2)^{\tau/2})\,.$$

$$\cdot\; \sigma^{\gamma_1}(iD_\sigma)^\gamma\, q(\sigma)\cdot q(\sigma)d\sigma\,.$$

$$\therefore\ p_\alpha(x,\xi) \in S^{m-(\tau-\delta)|\alpha|}_{\rho,\delta} \subset S^{m-(\rho-\delta)}_{\rho,\delta} \text{ if } |\alpha| \geq 2.$$

Thus we only have to check $p_\alpha(x,\zeta)$ for $|\alpha| = 0$ and 1.

Since q is even and $\int q(\sigma)^2 d\sigma = 1$,

$$\begin{aligned} p_o(x,\xi) &= \int p(x,\xi+\sigma(1+|\xi|^2)^{\tau/2})\, q(\sigma)^2 d\sigma \\ &= p(x,\xi) + p'(x,\xi), \end{aligned}$$

where

$$\begin{aligned} p'(x,\xi) &= 2(1+|\xi|^2)^\tau \sum_{|\gamma|=2} \int \frac{\sigma^\gamma}{\gamma!} \cdot \\ &\quad \cdot \left(\int_0^1 (1-t)(iD_\xi)^\gamma p(x+t\sigma(1+|\xi|^2)^{\tau/2})dt\right) q(\sigma)^2 d\sigma. \end{aligned}$$

We have used Taylor's formula to order 2, expanding in power of σ. (Conveniently, the linear term in σ integrates to zero, because q is even.) Hence $p'(x,\xi) \in S^{m-(\rho-\delta)}_{\rho,\delta}$.

To handle the case $|\alpha| = 1$, note that $\psi_{\alpha,\gamma,\gamma} \in S^{-1}_{1,0} \subset S^{-\rho}_{\rho,\delta}$, so we will get $p_\alpha(x,\xi) \in S^{m-(\rho-\delta)}_{\rho,\delta}$ from the fact that

$$\text{(1)} \qquad \psi_{\alpha,\gamma,0}(\xi) \int D_x^\alpha\, p(x,\xi+\sigma(1+|\xi|^2)^{\tau/2})(iD_\sigma)^\gamma q(\sigma)\cdot q(\sigma)d\sigma \in S^{m-(\rho-\delta)}_{\rho,\delta},$$

provided $|\alpha| = |\gamma| = 1$. This is proved by writing

$$\begin{aligned} p(x,\xi+\sigma(1+|\xi|^2)^{\tau/2}) &= p(x,\xi) + (1+|\xi|^2)^{\tau/2} \sum_{|\gamma|=1} \sigma^\gamma \cdot \\ &\quad \cdot \int_0^1 (iD_\xi)^\gamma p(x,\xi+t\sigma(1+|\xi|^2)^{\tau/2})dt. \end{aligned}$$

Since $\int D_\sigma^\gamma q(\sigma)\cdot q(\sigma) = 0$ (q is even); the first term gives in (1) no contribution, and the estimate goes through without difficulty. This completes the proof.

Corollary (sharp Gårding inequality): If $p \in S^m_{\rho,\delta}$, $0 \leq \delta < \rho \leq 1$, and if $p(x,\xi) \geq 0$, then

$$\mathrm{Re}(p(x,D)u,u) \geq -c\|u\|^2_{1/2(m-(\rho-\delta))}.$$

Proof: Since the Friedrich's symmetrization $b(D,x,D)$ is a positive operator and $b(D,x,D) - p(x,D) \in PS(m-(\rho-\delta),\rho,\delta)$, this inequality is immediate.

Exercise 4: Let M be a compact manifold, $K = K(t) \in PS(1,1,0)$ a smooth one parameter family of pseudo differential operators on M, and suppose $\sigma_K(x,\xi) + \sigma_K(x,\xi)^* \geq 0$. Discuss existence and uniqueness of solutions to the initial value problem

$$\frac{\partial}{\partial t} u = -Ku + f$$

$$u(0) = \varphi_0.$$

Exercise 5: If $K \in PS(1,1,0)$ on M and $|\sigma_K(x,\xi)| \leq C < \infty$, show that $K: L^2(M) \longrightarrow L^2(M)$.

REFERENCES

1. S. Agmon, Lectures on Elliptic Boundary Value Problems, van Nostrand, 1964.

2. Agmon, Douglis, and Nirenberg, "Estimates near the boundary for solution of elliptic differential equations satisfying general boundary conditions." Comm. Pure Appl. Math. 12(1959) 623-727.

3. M.S. Agranovich, "Boundary value problems for first order pseudodifferential operators." Russian Math. Surveys (1971) 59-126.

4. Boutet de Monvel, "Boundary problems for pseudo-differential operators." Acta. Math. 126(1971) 11-51.

5. ———————, "Comportenent d'un operateur pseudo differentiel sur une variété à bord." J. Anal. Math. 17(1966) 241-304.

6. A. Calderon and A. Zygmund, "Singular integral operators and differential equations." I. Amer. J. Math. 79(1957) 901-921, II. Amer. J. Math. 80(1958) 16-36.

7. ———————, "On the existence of certain singular integrals." Acta. Math. 88(1952) 85-139.

8. A. Calderon, "Algebras of singular integral operators." Proceedings of Symposia in Pure Math. X, Amer. Math. Soc. 1967.

9. L. Coburn, "Weyl's theorem for non normal operators." Mich. J. Math. 13(1966) 285-288.

10. H. Cordes, "Pseudo differential operators on a half-line." J. Math. Mech. 18(1969) 893-908.

11. ———————, "An algebra of singular integral operators with two symbol homorphisms" Bull. Amer. Math. Soc. 75(1969) 37-42.

12. H. Cordes and E. Hermann, "Gelfand theory of pseudo differential operators." Amer. J. Math. 90(1968) 681-717.

13. Courant and Hilbert, Methods of Mathematical Physics II. J. Wiley 1966.

14. Courant and Lax, "The propagation of discontinuities in wave motion." Proc. Nat. Acad. Sc. U.S.A. 42(1956) 872-876.

15. J. Dieudonne , Foundations of Modern Analysis, Academic Press, New York, 1964.

16. J. Dixmier, Les C* Algebres et Leurs Représentations, Gauthier-Villans. 1964.

17. J. Dugundji, Topology. Allyn and Bacon, New York 1966.

18. Duistermaat and Hörmander, "Fourier integral operators, II." Acta Math., to appear.

19. Yu. V. Egorov, "On canonical transformations of pseudo-differential operators." Uspeki Math. Nausk. 25(1969) 235-236.

20. Yu. V. Egorov, "On subelliptic pseudo-differential operators." Dokl. Akad. Nauck. S.S.S.R. 188(1969) 20-22. Soviet Math. Dokl. 10(1969) 1056-1059.

21. A. Freidman, Partial Differential Equations of Parabolic Type. Prentis Hall, Englewood Cliffs, N.H. 1964.

22. K. Friedrichs, Lectures on Pseudo Differential Operators. N.Y.U. Lecture Notes. 1968.

23. ______________, "The identity of weak and strong extensions of differential operators." Trans. Amer. Math. Soc. 55(1944) 132-151.

24. Friedrichs and Lax, "Boundary value problems for first order operators." Comm. Pure Appl. Math. 18(1965) 355-388.

25. P. Garabedian, Partial Differential Equations, J. Wiley. New York. 1964.

26. L. Gårding, "Dirichlet's problem for linear elliptic partial differential equations." Math. Scand. 1 (1953) 55-72.

27. ______________, "Solution directe de probleme de Cauchy pour les equations hyperboliques." Coll. Int. C.N.R.S. Nanoz. 1956. 71-90.

28. Goldstein, Classical Mechanics. Addison-Wesley. 1950.

29. E. Herman, "The symbol of the algebra of singular integral operators." J. Math. Mech. 15(1966) 147-156.

30. R. Hersch, "Mixed problems in serveral variables." J. Math. Mech. 12(1963) 317-334.

31. L. Hörmander, Linear Partial Differential Operators. Springer-Verlag, 1964.

32. ——————,"Pseudo-differential operators." Comm. Pure Appl. Math. 18(1965) 501-517.

33. ——————, "Pseudo-differential operators and non-elliptic boundary problems." Ann. of Math. 83(1966) 129-209.

34. ——————, "Pseudo-differential operators and hypoelliptic equations." Singular Integrals, Proc. Sym. Pure Math X, Amer. Math. Soc. 1967

35. ——————, "On the existence and the regularity of solutions of linear pseudo-differential equations." L'Enseigenent Math.

36. ——————, "Hypoelliptic second order differential equations." Acta. Math. 119(1967) 147-171.

37. ——————, "The spectral function of an elliptic operator." Acta Math. 121(1968) 193-218.

38. ——————, "The calculus of Fourier integral operators." Prospects in Mathematics,

39. ——————, "Fourier integral operators I." Acta. Math. 127(1971) 79-183.

40. T. Kato, Peturbation Theory for Linear Operators, Springer-Verlag 1966.

41. T. Kato, "Linear evolution equations of "hyperbolic" type." J.Fac.Sci. Tokyo, 17(1970) 241-258.

42. Kelley, General Topology Van Nostrand, New York, 1955.

43. J. Kohn, "Pseudo-differential operators and hypoellipticity." Berkeley PDE, to appear.

44. J. Kohn and L. Nirenberg, "An algeba of pseudo-differential operators." Comm. Pure Appl. Math. 18(1965) 269-305.

45. J. Kohn and L. Nirenberg, "Non-coercive boundary problems." Comm. Pure Appl. Math 18(1965).

46. H. Kreiss, "Initial boundary value problems for hyperbolic systems." Comm. Pure Appl. Math. 23(1970) 277-298.

47. H. Kumano-Go, "Pseudo-differential operators and the uniqueness in the Cauchy problem." Comm. Pure Appl. Math 22(1969) 73-129.

48. P. Lax, Lectures on Hyperbolic Equations. Stanford Lecture Notes. 1963.

49. ______________, "Asymptotic solutions of oscillatory initial value problems." Comm. Pure Appl. Math (1957).

50. ______________, "On Cauchy's problem for hyperbolic equations and the differentiability of solutions of elliptic equations." Comm. Pure Appl. Math. (1955) 615-633.

51. Lax and Nirenberg, "On stability for difference schemes; a sharp form of Gårding's inequality." Comm. Pure Appl. Math. 19(1966) 473-492.

52. Lax and Phillips, Scattering Theory. Academic Press, New York 1967.

53. ______________, "Scattering Theory". Rocky Mountain J. Math. 1(1971) 173-223.

54. Lions, Equations Differentielles Operationelles. Springer-Verlag. 1963.

55. Lions and Magenes, Problems aux Limites non Hamogenes et Leurs Applications Dunod, Paris, 1967.

56. D. Ludwig, "Exact and asymptotic solutions of the Cauchy problem." Comm. Pure Appl. Math 13(1960) 473-508.

57. McKean and Singer, "Curvature and eigenvalues of the Laplacian." Jorn. Diff. Geom. 1(1967) 43-69.

58. Mikhlin, Multidimensional Singular Integral Equations. Pergamon Press, New York. 1965.

59. C. Morrey, Multiple Integrals and the Calculus of Variations. Springer-Verlag 1966.

60. Muskhelishvili, Singular Integral Equations. P. Nordhoff, Groningen. 1953.

61. U. Neri, "Singular integral operators on manifolds." Singular Integrals. Proc. Symp. Pure Math X. Amer. Math. Soc. 1967.

62. L. Nirenberg and F. Treves, "On local solvability of linear partial differential equations." I and II. Comm. Pure Appl. Math. 23(1970) 1-38 and 459-510.

63. O. Oleinick and E. Radkevitch, "Second order equations with non-negative characteristic form." Math. Analysis 1969, Itogi Nauk, Moscow 1971. (English translation to appear).

64. Palais , ed., Seminar on the Atiyah-Singer Index Theorem, Princeton Annals of Math. Studies, 1963.

65. H. Poincare, Leçons de Mechanique Celeste. Paris, 1910.

66. J. Polking, "Boundary value problems for parabolic systems of partial differential equations." Singular Integrals, Proc. Symp. Pure Math., Amer. Math. Soc. 1967.

67. J. Ralston, "Deficiency indices of symmetric operators with elliptic boundary conditions." Comm. Pure Appl Math. 23(1970). 221-232.

68. J. Rauch, "L^2 is a continuable initial condition for Kreiss well posed problems." to appear.

69. J. Rauch and M. Taylor, "Penetrations into shadow regions and unique continuation properties in hyperbolic mixed problems." Indiana Jour. Math., to appear.

70. Richtmyer and Morton, Difference Methods for Initial-value Problems. J. Wiley, New York. 1967.

71. L. Sarason, "On weak and strong solutions of boundary value problems." Comm. Pure Appl. Math 15(1962) 237-288.

72. R. Seeley, "Singular integrals on compact manifolds." Amer. J. Math. 81 (1959) 686-690.

73. ——————, "Singular integrals and boundary value problems." Amer. J. Math. 88(1966) 781-809.

74. ——————, "Complex powers of an elliptic operator." Singular Integrals. Proc. Symp. Pure Math. Amer. Math. Soc. 1967.

75. D. Tartakoff, "Regularity of solutions to boundary value problems for first order systems." To appear.

76. M. Taylor, "Gelfand theory of pseudo-differential operators, and hypoelliptic operators." Trans. Amer. Math. Soc. 153(1971) 495-510.

77. F. Treves, Topological Vector Spaces, Distributions, and Kernels. Academic Press, New York. 1967.

78. ——————, "A new proof of the subelliptic estimates." Comm. Pure Appl. Math 24(1971) 71-115.

79. ——————, "Hypoelliptic partial differential equations of principal type with analytic coefficients." Comm. Pure Appl Math. 23(1970) 637-651.

80. ——————, Partial Differential Equations with Constant Coefficients. Gordon and Breach. New York 1967.

81. R. Vaillancourt, "A simple proof of the Lax-Nirenberg theorem." Comm. Pure Appl. Math. 23(1970) 151-163.

82. ——————, "Pseudo-translation operators." Thesis, N.Y.U. 1969.

83. Visik and Eskin, "Normally solvable problems for elliptic systems of equations in convolutions." Mat. Sboinik 74 (116) (1967) 326-356.

84. Yamaguti and Nogi, "An algebra of pseudo-difference schemes and its application." Publ. RIMS Vol. 3(1967) 151-166.

85. K. Yosida, Functional Analysis. Springer-Verlag. 1965.

86. R. Beals and C.Fefferman, "On local solvability of linear partial differential equations." Ann. of Math. (1973) 482-498.

87. ——————, "Spatially inhomogeneous pseudo-differential operators," Comm. Pure Appl. Math, to appear.

88. A. Calderon and R. Vaillancourt, "On the boundedness of pseudodifferential operators," J. Math. Soc. Japan, 23 (1971), 374-378.

89. J. Duistermaat, Fourier Integral Operators, Courant Institute Lecture Notes, 1973.

90. H. Kumano-Go, "Algebras of pseudodifferential operators," Jour. Fac. Sci. Tokyo, 17 (1970) 31-50.

91. L. Nirenberg, Lectures on Partial Differential Equations, Proc. Reg. Conf. at Texas Tech, 1972 (A.M.S. Publication.)

92. J. Rauch and M. Taylor, "Exponential decay of solutions to symmetric hyperbolic equations in bounded domains," Indiana J. Math. to appear June 1974.

Vol. 247: Lectures on Operator Algebras. Tulane University Ring and Operator Theory Year, 1970–1971. Volume II. XI, 786 pages. 1972. DM 40,-

Vol. 248: Lectures on the Applications of Sheaves to Ring Theory. Tulane University Ring and Operator Theory Year, 1970–1971. Volume III. VIII, 315 pages. 1971. DM 26,-

Vol. 249: Symposium on Algebraic Topology. Edited by P. J. Hilton. VII, 111 pages. 1971. DM 16,-

Vol. 250: B. Jónsson, Topics in Universal Algebra. VI, 220 pages. 1972. DM 20,-

Vol. 251: The Theory of Arithmetic Functions. Edited by A. A. Gioia and D. L. Goldsmith VI, 287 pages. 1972. DM 24,-

Vol. 252: D. A. Stone, Stratified Polyhedra. IX, 193 pages. 1972. DM 18,-

Vol. 253: V. Komkov, Optimal Control Theory for the Damping of Vibrations of Simple Elastic Systems. V, 240 pages. 1972. DM 20,-

Vol. 254: C. U. Jensen, Les Foncteurs Dérivés de $\varprojlim$ et leurs Applications en Théorie des Modules. V, 103 pages. 1972. DM 16,-

Vol. 255: Conference in Mathematical Logic – London '70. Edited by W. Hodges. VIII, 351 pages. 1972. DM 26,-

Vol. 256: C. A. Berenstein and M. A. Dostal, Analytically Uniform Spaces and their Applications to Convolution Equations. VII, 130 pages. 1972. DM 16,-

Vol. 257: R. B. Holmes, A Course on Optimization and Best Approximation. VIII, 233 pages. 1972. DM 20,-

Vol. 258: Séminaire de Probabilités VI. Edited by P. A. Meyer. VI, 253 pages. 1972. DM 22,-

Vol. 259: N. Moulis, Structures de Fredholm sur les Variétés Hilbertiennes. V, 123 pages. 1972. DM 16,-

Vol. 260: R. Godement and H. Jacquet, Zeta Functions of Simple Algebras. IX, 188 pages. 1972. DM 18,-

Vol. 261: A. Guichardet, Symmetric Hilbert Spaces and Related Topics. V, 197 pages. 1972. DM 18,-

Vol. 262: H. G. Zimmer, Computational Problems, Methods, and Results in Algebraic Number Theory. V, 103 pages. 1972. DM 16,-

Vol. 263: T. Parthasarathy, Selection Theorems and their Applications. VII, 101 pages. 1972. DM 16,-

Vol. 264: W. Messing, The Crystals Associated to Barsotti-Tate Groups: With Applications to Abelian Schemes. III, 190 pages. 1972. DM 18,-

Vol. 265: N. Saavedra Rivano, Catégories Tannakiennes. II, 418 pages. 1972. DM 26,-

Vol. 266: Conference on Harmonic Analysis. Edited by D. Gulick and R. L. Lipsman. VI, 323 pages. 1972. DM 24,-

Vol. 267: Numerische Lösung nichtlinearer partieller Differential- und Integro-Differentialgleichungen. Herausgegeben von R. Ansorge und W. Törnig, VI, 339 Seiten. 1972. DM 26,-

Vol. 268: C. G. Simader, On Dirichlet's Boundary Value Problem. IV, 238 pages. 1972. DM 20,-

Vol. 269: Théorie des Topos et Cohomologie Etale des Schémas. (SGA 4). Dirigé par M. Artin, A. Grothendieck et J. L. Verdier. XIX, 525 pages. 1972. DM 50,-

Vol. 270: Théorie des Topos et Cohomologie Etale des Schémas. Tome 2. (SGA 4). Dirigé par M. Artin, A. Grothendieck et J. L. Verdier. V, 418 pages. 1972. DM 50,-

Vol. 271: J. P. May, The Geometry of Iterated Loop Spaces. IX, 175 pages. 1972. DM 18,-

Vol. 272: K. R. Parthasarathy and K. Schmidt, Positive Definite Kernels, Continuous Tensor Products, and Central Limit Theorems of Probability Theory. VI, 107 pages. 1972. DM 16,-

Vol. 273: U. Seip, Kompakt erzeugte Vektorraume und Analysis. IX, 119 Seiten. 1972. DM 16,-

Vol. 274: Toposes, Algebraic Geometry and Logic. Edited by. F. W. Lawvere. VI, 189 pages. 1972. DM 18,-

Vol. 275: Séminaire Pierre Lelong (Analyse) Année 1970–1971. VI, 181 pages. 1972. DM 18,-

Vol. 276: A. Borel, Représentations de Groupes Localement Compacts. V, 98 pages. 1972. DM 16,-

Vol. 277: Séminaire Banach. Edité par C. Houzel. VII, 229 pages. 1972. DM 20,-

Vol. 278: H. Jacquet, Automorphic Forms on GL(2). Part II. XIII, 142 pages. 1972. DM 16,-

Vol. 279: R. Bott, S. Gitler and I. M. James, Lectures on Algebraic and Differential Topology. V, 174 pages. 1972. DM 18,-

Vol. 280: Conference on the Theory of Ordinary and Partial Differential Equations. Edited by W. N. Everitt and B. D. Sleeman. XV, 367 pages. 1972. DM 26,-

Vol. 281: Coherence in Categories. Edited by S. Mac Lane. VII, 235 pages. 1972. DM 20,-

Vol. 282: W. Klingenberg und P. Flaschel, Riemannsche Hilbertmannigfaltigkeiten. Periodische Geodätische. VII, 211 Seiten. 1972. DM 20,-

Vol. 283: L. Illusie, Complexe Cotangent et Déformations II. VII, 304 pages. 1972. DM 24,-

Vol. 284: P. A. Meyer, Martingales and Stochastic Integrals I. VI, 89 pages. 1972. DM 16,-

Vol. 285: P. de la Harpe, Classical Banach-Lie Algebras and Banach-Lie Groups of Operators in Hilbert Space. III, 160 pages. 1972. DM 16,-

Vol. 286: S. Murakami, On Automorphisms of Siegel Domains. V, 95 pages. 1972. DM 16,-

Vol. 287: Hyperfunctions and Pseudo-Differential Equations. Edited by H. Komatsu. VII, 529 pages. 1973. DM 36,-

Vol. 288: Groupes de Monodromie en Géométrie Algébrique. (SGA 7 I). Dirigé par A. Grothendieck. IX, 523 pages. 1972. DM 50,-

Vol. 289: B. Fuglede, Finely Harmonic Functions. III, 188. 1972. DM 18,-

Vol. 290: D. B. Zagier, Equivariant Pontrjagin Classes and Applications to Orbit Spaces. IX, 130 pages. 1972. DM 16,-

Vol. 291: P. Orlik, Seifert Manifolds. VIII, 155 pages. 1972. DM 16,-

Vol. 292: W. D. Wallis, A. P. Street and J. S. Wallis, Combinatorics: Room Squares, Sum-Free Sets, Hadamard Matrices. V, 508 pages. 1972. DM 50,-

Vol. 293: R. A. DeVore, The Approximation of Continuous Functions by Positive Linear Operators. VIII, 289 pages. 1972. DM 24,-

Vol. 294: Stability of Stochastic Dynamical Systems. Edited by R. F. Curtain. IX, 332 pages. 1972. DM 26,-

Vol. 295: C. Dellacherie, Ensembles Analytiques, Capacités, Mesures de Hausdorff. XII, 123 pages. 1972. DM 16,-

Vol. 296: Probability and Information Theory II. Edited by M. Behara, K. Krickeberg and J. Wolfowitz. V, 223 pages. 1973. DM 20,-

Vol. 297: J. Garnett, Analytic Capacity and Measure. IV, 138 pages. 1972. DM 16,-

Vol. 298: Proceedings of the Second Conference on Compact Transformation Groups. Part 1. XIII, 453 pages. 1972. DM 32,-

Vol. 299: Proceedings of the Second Conference on Compact Transformation Groups. Part 2. XIV, 327 pages. 1972. DM 26,-

Vol. 300: P. Eymard, Moyennes Invariantes et Représentations Unitaires. II, 113 pages. 1972. DM 16,-

Vol. 301: F. Pittnauer, Vorlesungen uber asymptotische Reihen. VI, 186 Seiten. 1972. DM 18,-

Vol. 302: M. Demazure, Lectures on p-Divisible Groups. V, 98 pages. 1972. DM 16,-

Vol. 303: Graph Theory and Applications. Edited by Y. Alavi, D. R. Lick and A. T. White. IX, 329 pages. 1972. DM 26,-

Vol. 304: A. K. Bousfield and D. M. Kan, Homotopy Limits, Completions and Localizations. V, 348 pages. 1972. DM 26,-

Vol. 305: Théorie des Topos et Cohomologie Etale des Schémas. Tome 3. (SGA 4). Dirigé par M. Artin, A. Grothendieck et J. L. Verdier. VI, 640 pages. 1973. DM 50,-

Vol. 306: H. Luckhardt, Extensional Gödel Functional Interpretation. VI, 161 pages. 1973. DM 18,-

Vol. 307: J. L. Bretagnolle, S. D. Chatterji et P.-A. Meyer, Ecole d'été de Probabilités: Processus Stochastiques. VI, 198 pages. 1973. DM 20,-

Vol. 308: D. Knutson, λ-Rings and the Representation Theory of the Symmetric Group. IV, 203 pages. 1973. DM 20,-

Vol. 309: D. H. Sattinger, Topics in Stability and Bifurcation Theory. VI, 190 pages. 1973. DM 18,-

Vol. 310: B. Iversen, Generic Local Structure of the Morphisms in Commutative Algebra. IV, 108 pages 1973. DM 16,-

Vol. 311: Conference on Commutative Algebra. Edited by J. W. Brewer and E. A. Rutter. VII, 251 pages. 1973 DM 22,-

Vol. 312: Symposium on Ordinary Differential Equations. Edited by W. A. Harris, Jr. and Y. Sibuya. VIII, 204 pages. 1973. DM 22,-

Vol. 313: K. Jörgens and J. Weidmann, Spectral Properties of Hamiltonian Operators. III, 140 pages 1973 DM 16,-

Vol. 314: M. Deuring, Lectures on the Theory of Algebraic Functions of One Variable. VI, 151 pages. 1973. DM 16,-

Vol. 315: K. Bichteler, Integration Theory (with Special Attention to Vector Measures) VI, 357 pages. 1973. DM 26,-

Vol. 316: Symposium on Non-Well-Posed Problems and Logarithmic Convexity. Edited by R. J. Knops. V, 176 pages. 1973. DM 18,-

Vol. 317: Séminaire Bourbaki – vol. 1971/72. Exposés 400–417. IV, 361 pages. 1973. DM 26,-

Vol. 318: Recent Advances in Topological Dynamics. Edited by A. Beck, VIII, 285 pages. 1973. DM 24,-

Vol. 319: Conference on Group Theory. Edited by R. W. Gatterdam and K. W. Weston. V, 188 pages. 1973 DM 18,-

Vol. 320: Modular Functions of One Variable I. Edited by W. Kuyk. V, 195 pages. 1973. DM 18,-

Vol. 321: Séminaire de Probabilités VII. Edité par P. A. Meyer. VI, 322 pages. 1973. DM 26,-

Vol. 322: Nonlinear Problems in the Physical Sciences and Biology. Edited by I. Stakgold, D. D. Joseph and D. H. Sattinger. VIII, 357 pages. 1973. DM 26,-

Vol. 323: J. L. Lions, Perturbations Singulières dans les Problèmes aux Limites et en Contrôle Optimal. XII, 645 pages. 1973. DM 42,-

Vol. 324: K. Kreith, Oscillation Theory. VI, 109 pages. 1973. DM 16,-

Vol. 325. Ch.-Ch. Chou, La Transformation de Fourier Complexe et L'Equation de Convolution. IX, 137 pages. 1973. DM 16,-

Vol. 326: A. Robert, Elliptic Curves. VIII, 264 pages. 1973. DM 22,-

Vol. 327: E. Matlis, 1-Dimensional Cohen-Macaulay Rings. XII, 157 pages. 1973. DM 18,-

Vol. 328: J. R. Büchi and D. Siefkes, The Monadic Second Order Theory of All Countable Ordinals. VI, 217 pages. 1973. DM 20,-

Vol. 329: W. Trebels, Multipliers for (C, α)-Bounded Fourier Expansions in Banach Spaces and Approximation Theory. VII, 103 pages. 1973. DM 16,-

Vol. 330: Proceedings of the Second Japan-USSR Symposium on Probability Theory. Edited by G. Maruyama and Yu. V. Prokhorov. VI, 550 pages. 1973. DM 36,-

Vol. 331: Summer School on Topological Vector Spaces. Edited by L. Waelbroeck. VI, 226 pages. 1973. DM 20,-

Vol. 332: Séminaire Pierre Lelong (Analyse) Année 1971-1972. V, 131 pages. 1973. DM 16,-

Vol. 333: Numerische, insbesondere approximationstheoretische Behandlung von Funktionalgleichungen. Herausgegeben von R. Ansorge und W. Törnig. VI, 296 Seiten. 1973. DM 24,-

Vol. 334: F. Schweiger, The Metrical Theory of Jacobi-Perron Algorithm. V, 111 pages. 1973. DM 16,-

Vol. 335: H. Huck, R. Roitzsch, U. Simon, W. Vortisch, R. Walden, B. Wegner und W. Wendland, Beweismethoden der Differentialgeometrie im Großen. IX, 159 Seiten. 1973. DM 18,-

Vol. 336: L'Analyse Harmonique dans le Domaine Complexe. Edité par E. J. Akutowicz. VIII, 169 pages. 1973. DM 18,-

Vol. 337: Cambridge Summer School in Mathematical Logic. Edited by A. R. D. Mathias and H. Rogers. IX, 660 pages. 1973. DM 42,-

Vol. 338: J. Lindenstrauss and L. Tzafriri, Classical Banach Spaces. IX, 243 pages. 1973. DM 22,-

Vol. 339: G. Kempf, F. Knudsen, D. Mumford and B. Saint-Donat, Toroidal Embeddings I. VIII, 209 pages. 1973. DM 20,-

Vol. 340: Groupes de Monodromie en Géométrie Algébrique. (SGA 7 II). Par P. Deligne et N. Katz. X, 438 pages. 1973. DM 40,-

Vol. 341: Algebraic K-Theory I, Higher K-Theories. Edited by H. Bass. XV, 335 pages. 1973. DM 26,-

Vol. 342: Algebraic K-Theory II, "Classical" Algebraic K-Theory, and Connections with Arithmetic. Edited by H. Bass. XV, 527 pages. 1973. DM 36,-

Vol. 343: Algebraic K-Theory III, Hermitian K-Theory and Geometric Applications. Edited by H. Bass. XV, 572 pages. 1973. DM 38,-

Vol. 344: A. S. Troelstra (Editor), Metamathematical Investigation of Intuitionistic Arithmetic and Analysis. XVII, 485 pages. 1973. DM 34,-

Vol. 345: Proceedings of a Conference on Operator Theory. Edited by P. A. Fillmore. VI, 228 pages. 1973 DM 20,-

Vol. 346: Fučík et al., Spectral Analysis of Nonlinear Operators. II, 287 pages. 1973. DM 26,-

Vol. 347: J. M. Boardman and R. M. Vogt, Homotopy Invariant Algebraic Structures on Topological Spaces. X, 257 pages. 1973. DM 22,-

Vol. 348: A. M. Mathai and R. K. Saxena, Generalized Hypergeometric Functions with Applications in Statistics and Physical Sciences. VII, 314 pages. 1973. DM 26,-

Vol. 349: Modular Functions of One Variable II. Edited by W. Kuyk and P. Deligne. V, 598 pages. 1973. DM 38,-

Vol. 350: Modular Functions of One Variable III. Edited by W. Kuyk and J.-P. Serre. V, 350 pages. 1973. DM 26,-

Vol. 351: H. Tachikawa, Quasi-Frobenius Rings and Generalizations. XI, 172 pages. 1973. DM 18,-

Vol. 352: J. D. Fay, Theta Functions on Riemann Surfaces. V, 137 pages. 1973. DM 16,-

Vol. 353: Proceedings of the Conference. on Orders, Group Rings and Related Topics. Organized by J. S. Hsia, M. L. Madan and T. G. Ralley. X, 224 pages. 1973. DM 20,-

Vol. 354: K. J. Devlin, Aspects of Constructibility. XII, 240 pages. 1973. DM 22,-

Vol. 355: M. Sion, A Theory of Semigroup Valued Measures. V, 140 pages. 1973. DM 16,-

Vol. 356: W. L. J. van der Kallen, Infinitesimally Central-Extensions of Chevalley Groups. VII, 147 pages. 1973. DM 16,-

Vol. 357: W. Borho, P. Gabriel und R. Rentschler, Primideale in Einhullenden auflösbarer Lie-Algebren. V, 182 Seiten. 1973. DM 18,-

Vol. 358: F. L. Williams, Tensor Products of Principal Series Representations. VI, 132 pages. 1973. DM 16,-

Vol. 359: U. Stammbach, Homology in Group Theory. VIII, 183 pages. 1973. DM 18,-

Vol. 360. W. J. Padgett and R. L. Taylor, Laws of Large Numbers for Normed Linear Spaces and Certain Fréchet Spaces. VI, 111 pages. 1973. DM 16,-

Vol. 361: J. W. Schutz, Foundations of Special Relativity: Kinematic Axioms for Minkowski Space Time. XX, 314 pages. 1973. DM 26,-

Vol. 362: Proceedings of the Conference on Numerical Solution of Ordinary Differential Equations. Edited by D. Bettis. VIII, 490 pages. 1974. DM 34,-

Vol. 363: Conference on the Numerical Solution of Differential Equations. Edited by G. A. Watson. IX, 221 pages. 1974. DM 20,-

Vol. 364: Proceedings on Infinite Dimensional Holomorphy. Edited by T. L. Hayden and T. J. Suffridge. VII, 212 pages. 1974. DM 20,-

Vol. 365: R. P. Gilbert, Constructive Methods for Elliptic Equations. VII, 397 pages. 1974. DM 26,-

Vol. 366: R. Steinberg, Conjugacy Classes in Algebraic Groups (Notes by V. V. Deodhar). VI, 159 pages. 1974. DM 18,-

Vol. 367: K. Langmann und W. Lütkebohmert, Cousinverteilungen und Fortsetzungssätze. VI, 151 Seiten. 1974. DM 16,-

Vol. 368: R. J. Milgram, Unstable Homotopy from the Stable Point of View. V, 109 pages. 1974. DM 16,-

Vol. 369: Victoria Symposium on Nonstandard Analysis. Edited by A. Hurd and P. Loeb. XVIII, 339 pages. 1974. DM 26,-

Vol. 370: B. Mazur and W. Messing, Universal Extensions and One Dimensional Crystalline Cohomology. VII, 134 pages. 1974. DM 16,-